COASTWISE

For Paula

COASTWISE

Understanding Britain's Shoreline

Peter Firstbrook

FERNHURST
BOOKS

First published in 2021 by Fernhurst Books Limited

The Windmill, Mill Lane, Harbury, Leamington Spa,
Warwickshire. CV33 9HP, UK
Tel: +44 (0) 1926 337488 | www.fernhurstbooks.com

A catalogue record for this book is available from the British
Library
ISBN 978-1-912621-40-8

Cover photograph Flamborough Head; © TimHill / Pixabay

Designed & typeset by Daniel Stephen
Printed in Czech Republic by Finidr

CONTENTS

A common sea pup in the North Sea.

FOREWORD

This subtly passionate book is the perfect companion to anyone with an interest in the coast and, given our proximity to water and reliance on what it delivers us, that should be all of us.

In my work for *Countryfile* and *Costing the Earth* I've been lucky enough to enjoy many of the nation's coasts and they always come with a 'buzz'. It's why as children on a trip to the seaside you strive to be the first to see the sea. My broadcasting work takes me there so often because they are a source of such good stories, an upwelling of issues: flooding, fishing, seaweed farming, smuggling and so many more. All to be found in these pages. As a sailor, scientist and programme maker Peter Firstbrook is clearly smitten by the bit of this country where solid meets liquid. And I get that. It is a dynamic strip of geography. A place of constant change and thrilling unpredictability. Benign golden sand or ship-shattering rock, computer-controlled container ports or snipe on a salt marsh, the author warms to them all and his feelings are infectious.

The breadth of this book is staggering. Peter has taken the whole of Britain and Northern Ireland and, a bit like a parent rightly unwilling to choose their favourite child, has decided to spread the love. You can learn about Roman naval harbours on the Cumbrian coast or kite surfing in Thurso.

The sections of the book loosely follow an epic chronology of the forces affecting our coast – geology with tales of hard basalt and soft sandstone; geography and the effects of erosion, currents and weather; biology with abundant sealife; and then humanity with our trade, fishing and fun. This all anchors the reader with the formative facts of the shoreline.

Coastwise is unafraid of a little light science and much the better for it. Knowledge helps us get so much more out of the coast. Understanding why it looks like it does and why animals and humans thrive in certain spots underpins our enjoyment. Want to know why those corrugations appear on a sandy beach? The answer's in here. Want to know how to identify some edible seaweeds? The answer's in here. Want to know where best to spot an Atlantic white-sided dolphin? You get the picture.

The author wants us to experience the edge of the ocean first-hand and preferably immersed. The book comes equipped not only with a guide to where to discover the best of the British coast but also how to behave safely when you get there. There are tips on navigation buoys, how to avoid or survive a rip current and even how to conserve body heat in cold water with the 'Heat Escape Lessening Posture' – HELP – a piece of learning for this reader for sure.

If you'd rather avoid all that perilous wet stuff you can relax with a dry gin and admire the crashing waves from aboard one of the twelve most spectacular coastal railway routes also listed in this book.

These pages are a practical love letter to Britain's waterfront and no seaside holiday home should be without them.

Tom Heap, *Countryfile* Presenter

AUTHOR'S NOTE

One of my earliest recollections as a young boy is a family holiday to Cullercoats on the Northumberland coast. It was post-war Britain, and I have two particularly strong memories. The first was a dreary boarding house with peeling paper on the walls and a wilting aspidistra in the hall. This could not have been in stronger contrast to the aquarium on the seafront. Here was an exotic, weird and wonderful world of vibrant life and colour. I was mesmerised by what was living just off the beach.

That memory has never left me. I studied and researched marine sciences for nearly a decade, joined the BBC and made documentaries about oceanography, and spent much of my life sailing around Britain – and further afield when I could; I surfed as a student, and went scuba diving to see what life was like under the waves. In all this time, I never lost the excitement of that first eye-opening experience.

Here you are dazzled by the spectacular blue and orange cuckoo wrasse – every bit as colourful as any tropical reef fish, entranced by a dolphin breaking the surface, or enthralled by a puffin flying home with a mouthful of sand eels for its chick. You can gaze in awe as an Atlantic storm pounds a Cornish headland, or watch fascinated as a shoal of mackerel turn a Scottish sea loch into a feeding maelstrom as they force small prey to the surface – much to the delight of the hungry gannets circling above, who help themselves to an opportunistic take-away.

These are just some of the delights that our coastline has to offer. But what intrigues me even more is how all these events fit into a unique, interactive system. Tidal currents move plankton around, which are food for hungry fish; overhead, seabirds help themselves to an easy meal before returning to nest in the cliffs, which are formed by constant wave erosion over millions of years. These eroded cliffs are the source of the sand which is carried off by these very same currents to another part of the coastline to form beaches, bays and estuaries – all habitats for more marine life. And so the cycle continues…

The topics covered here are so broad that this book cannot be all-inclusive. Instead, it seeks to excite, inspire, and highlight the links between geology, oceanography (Part One), and biology (Part Two). For those readers interested to know more, there are excellent field guides available covering all areas of the marine world, whether your interest is in rocks, plants, seabirds or animals. Part Three looks at how our coastal nation has shaped our history, and how we have shaped our coastline. Finally, Part Four is a guide to where you can experience some of the features covered in this book.

This is our maritime heritage, and it is here to be enjoyed and appreciate by all.

Peter Firstbrook
September 2021

INTRODUCTION

THIS PRECIOUS STONE SET IN THE SILVER SEA

This precious stone set in the silver sea,
Which serves it in the office of a wall
Or as a moat defensive to a house...

From: *Richard II*
William Shakespeare

Swyre beach in Dorset, looking west towards Bat's Head.

If any organisation can be certain of the length of the British coastline, then it must be the Ordnance Survey. They have come up with a figure of 17,820km (11,070 miles) for Great Britain; add the length of the coastline of Northern Ireland, and the total for the United Kingdom is greater than the shorelines of Germany, France, Spain and Portugal combined.

Map makers have also identified 6,289 islands in Britain, most of them in Scotland. Admittedly only 803 are large enough to be properly digitised for map-making (the rest are recorded as 'point features'), but you get the idea. Greece, by comparison, claims to have 6,000 islands and islets. Clearly, we are blessed with an astonishingly diverse coastline, and it comes wrapped up in a fascinating history.

The Green Bridge of Wales in Pembrokeshire; a sea arch carved out of 350-million-year-old limestone.

Nobody in Britain lives more than 120km (75 miles) from the ocean, and our closeness to the coast has shaped the lives of us all; we are an island nation, and the sea is in our blood. We have inherited a coastline that ranges from the granite cliffs of Cornwall rising like cathedrals from the shore, to the sandbanks of Morecombe Bay, spread as flat as a pool table for as far as the eye can see; from the white-knuckle surfing beaches of the Gower peninsula in Wales, to the rocky ice-scooped sea lochs of Scotland. We share a coastline which offers a fascinating diversity for walkers, sailors, fishermen, paddlers and indeed anyone who simply wants to sit on a beach and appreciate the wonderful world around them.

Yet nothing is ever static, and our coastline is the product of painfully slow change over hundreds of millions of years. Great continents have moved and collided, mountain ranges have risen, volcanoes have erupted, ice caps have advanced across the surface, tropical forests have covered the land and dinosaurs have roamed the countryside. Superimposed on this agonisingly slow change is the relentless daily bombardment of the shoreline by the power of waves, wind and currents. The result is a coastal landscape which is more diverse than anywhere else on the planet.

WIND, WAVES, TIDES & CURRENTS

The British islands are wrapped around the north-western edge of the European continent, and they are more exposed to the power of wind and waves coming in from the Atlantic Ocean than most of our continental neighbours. Watch any television weather forecast, and the likelihood is the presenter will be showing 'the weather' coming in from the west and south-west. This is the prevailing wind direction in Britain, and it affects the way our coastline is eroded and how the **sediments** (sands, muds, and stones) are moved along the coastline.

The Needles on the Isle of Wight under the stormy sky of a severe force 9 gale. It is during storm conditions like this that most coastal erosion occurs.

The size of waves, combined with the power of tidal currents, create a **high-energy environment** from the West Country, through Wales, to the west coast of Ireland and Scotland. Here, large waves driven by **onshore** prevailing winds have the power to constantly shape and re-form the shoreline, creating a region dominated by **erosion**.

On the east coast of Scotland and south into the North Sea, the prevailing winds are **offshore**, and the waves are generally more subdued, creating a **low-energy environment**. As a result, much of the coastline here is a low-lying region of mainly **deposition**.

In the real world, of course, things are never quite so clear cut and there is always a mixture of erosion and deposition along any stretch of shoreline. Even the most rugged coasts of Scotland, Wales and the West Country have sand and shingle bays set between rocky headlands.

Rhossili beach in Wales is typical of much of the west coast of the British Isles where rocky headlands are interspersed with magnificent sandy beaches.

What these contrasting coastlines offer is a wide variety of habitats to support marine wildlife. We have more than 300 species of fish in our coastal waters, 26 species of marine mammals, 25 species of native seabirds and hundreds of different crustaceans, molluscs and marine plants – each adapted to survive in their particular habitat, whether it be the clifftops of a Hebridean island or a salt marsh in Norfolk.

Morecambe Bay, Lancashire, is the largest expanse of inter-tidal mudflats and sand in the United Kingdom and covers a total area of 310sq.km (120sq. miles).

THE ICE AGES

The other major influence which has shaped our coastline has been changing sea levels. Over the last century, the warming of our planet due to burning fossil fuels has caused sea levels to rise at an unprecedented rate. But sea level has always fluctuated over millions of years from natural causes, and this too has left its mark on the British coastline.

The period from 2.5 million years ago to around 11,700 years ago is called the **Quaternary** period by geologists, during which time the planet cooled; vast ice caps formed over the poles several kilometres thick, and they locked up so much water globally that sea level fell by as much as 130m (430ft) below today's level.

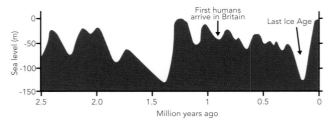

Changes in global sea level in the last 2.5 million years.

As the planet cooled and moved into a **glacial period**, sea level fell, and the British coastline expanded. For thousands of years the region was no longer an island but connected to continental Europe by a land bridge across the shallow North Sea – this region has been called **Doggerland**. We are beginning to learn a lot more about this period of history because, in May 2013, archaeologists working on Happisburgh beach in Norfolk made an extraordinary discovery: human footprints made around 900,000 years ago – the oldest human footprints found anywhere in the world outside of Africa.

These impressions were found at low tide in sediment that was partially covered by beach sand, and storms had washed away the sediment and exposed the footprints. Because the sediment was soft and the footprints were found below high water, the incoming tide soon began to erode the evidence. The team worked frantically at low water, often in the pouring rain, and were able to record 3D images of all the markings; within two weeks, the footprints were lost.

These prehistoric people were early hunter-gatherers known as *Homo antecessor*, and they came from continental Europe, and migrated through Doggerland and settled in what is now East Anglia. Archaeologists also found the flint tools they left behind as they walked along the mudflats of a long-lost estuary. This was an important and exciting discovery, and evidence that early humans occupied northern Europe at least 350,000 years *earlier* than was previously thought.

The coastline of the British Isles looked very different during the last glacial period. Britain was joined to continental Europe by a land bridge called Doggerland, which allowed the migration of the first humans, around 900,000 years ago.

Ever since that very early migration into the British islands, people have continued to leave their mark on our coastline. In Poole harbour, for example, archaeologists have found evidence of ancient wooden piling going back more than 2,000 years, thought to be a quayside used by Iron Age traders who sailed from France to buy pottery, jewellery and other items made locally in Dorset. The Iron Age Celts also built fortified strongholds around our coastline, many of which can still be visited.

MORE RECENT HISTORY

In 55 BC and again in AD 43, the Romans invaded *Britannia*. They had a rocky start at first and could not handle the vagaries of British tides. But once they got their armies ashore and settled down, they built ports, towns and lighthouses around the coastline of England. The Romans were followed by the Saxons, the Vikings, then the Normans – all of them leaving a coastal legacy that we can still see today.

The origins of the Tower of London, for example, date back to 1066. It was built by William the Conqueror to protect London from attack from the sea (as well as keeping the rebellious citizens of London in line). Shakespeare's *Richard II* even claimed the whole country was a defensive structure, a '*fortress built by Nature for herself, Against infection and the hand of war...*' From Medieval castles to the Second World War concrete pillboxes, our coastline is littered with our attempts – some more successful than others – to prevent invasion.

The early Christians built monasteries and abbeys along the coast for very practical reasons – the sea gave them easy access to spread their gospel, and the coastline offered a bountiful supply of food. Whitby Abbey is one of the oldest Christian buildings in the country, and has stood solid for nearly 1,500 years, offering solace to believers and creative inspiration to Bram Stoker, the author of *Dracula*.

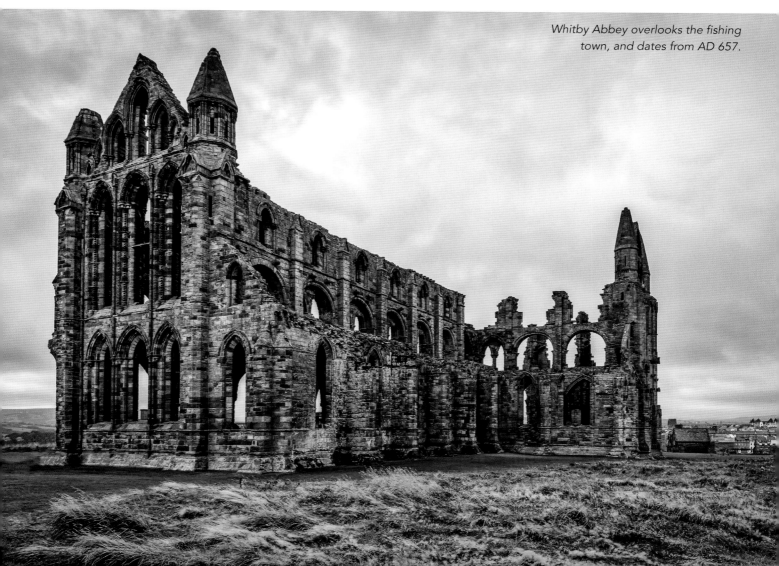

Whitby Abbey overlooks the fishing town, and dates from AD 657.

Nor has religious worship been restricted to grand buildings. At Beer on the south coast of Devon is a man-made underground complex about 1.6km (1 mile) west of the village. Roman artefacts have been found in the caverns, which suggests that limestone quarrying there goes back at least 2,000 years. After the Reformation, the caves were allegedly used as a meeting place for Catholics to secretly worship, safe from the prying eyes of the newly formed Protestant Church. By the nineteenth century, the caves had a more temporal use, and were appropriated by local smugglers to store contraband. The quarry is now part of the Jurassic Coast World Heritage site.

Over the centuries, industry has also left its mark on the British coastline. Mining in Devon and Cornwall began more than 4,000 years ago during the early Bronze Age, and some archaeologists believe Phoenician metal traders even sailed from the eastern Mediterranean in search of tin. We know for certain that the Greek geographer, trader and explorer, Pytheas of Massalia, sailed to Britain in 325 BC and found a flourishing trade in this metal (which is essential to combine with copper to make bronze). The abundance of tin was also one of the main reasons for the Roman invasion of the British Isles, and the invaders also mined slate as early as AD 77 to create the roof of the coastal fort of *Segontium* – or Carnarvon as it is better known today.

The engine houses at the Botallack submarine mine in west Cornwall, part of the UNESCO World Heritage Site since 2006. Early records can be traced back to the 1500s, but there is evidence of mining here during the Roman period, and even as far back as the Bronze Age.

With the Industrial Revolution, coal and iron were extracted in vast quantities, and many of these mines have left a lasting indentation along our coastline. Today, we are experiencing a new Industrial Revolution, and our coastline is now dotted with nuclear power stations, offshore windfarms, offshore gas and oil rigs, and a few (but not enough) generating stations using tidal power.

The London Array wind farm, one of the largest in the Thames estuary. There are now more than 30 offshore wind farms around Britain.

The British coastline has a history and heritage that has no equal. The shape and form of the coast can be traced back millions of years; superimposed on this is evidence of human occupation which goes back thousands of years.

Coastwise offers a fresh insight on this wonderful landscape, by looking at the British coastline as a complex but integrated system and opens a new window on understanding the fascinating shoreline around our islands.

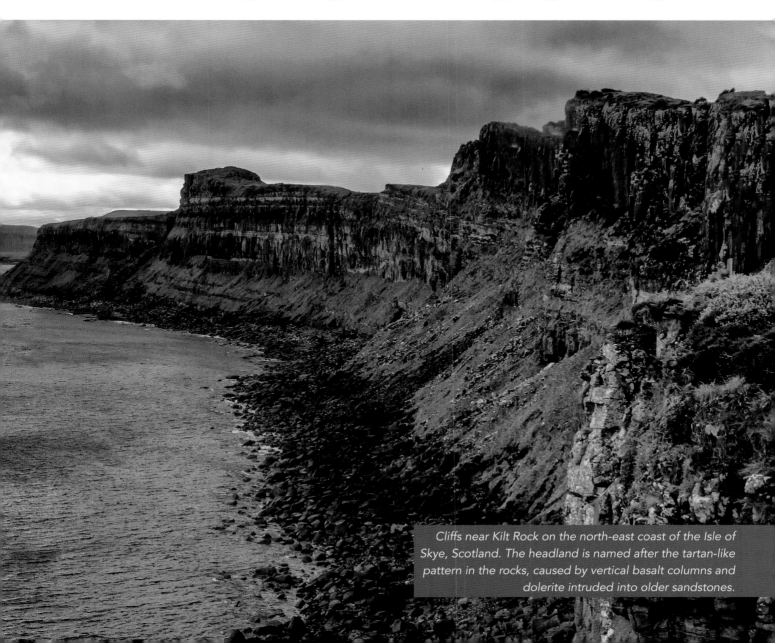

PART 1

THE PHYSICAL COAST

Cliffs near Kilt Rock on the north-east coast of the Isle of Skye, Scotland. The headland is named after the tartan-like pattern in the rocks, caused by vertical basalt columns and dolerite intruded into older sandstones.

CHAPTER 1

HEADLANDS & CLIFFS

150-million-year-old limestone cliffs on the Jurassic Coast, at Bridport in Dorset. This was designated Britain's first coastal UNESCO World Heritage Site.

Rackwick Bay on the Island of Hoy in the Orkney Islands is one of the most remote and ruggedly beautiful parts of the British Isles. The beach here is a mixture of fine sand and large boulders and provides a home to a variety of plants and birds that survive on a very exposed site, facing 3,500km (2,175 miles) of open Atlantic Ocean. From the beach you can look up at St John's Head, 335m (1,128ft) high and the tallest vertical sea cliff in Britain; nearby is the Old Man of Hoy, the tallest sea stack in the country.

An aquatint by William Daniell of the Old Man of Hoy (left) showing two stubby legs and an arch, c.1817. Shortly after the painting was completed, one of the legs crumpled during a storm and the feature partially collapsed. Today (right), the Old Man is a single stack.

The cliffs here are made from Old Red Sandstone laid down 370 million years ago in a dry desert, long before dinosaurs roamed the Earth. Gales blow for more than 30 days a year, creating high-energy waves which roll in from the North Atlantic. These ancient cliffs are no match for the pounding they receive from the ocean, and the fascinating story of the Old Man of Hoy offers a poignant example of how even the most resilient parts of our coastline can change over the years.

Less than 300 years ago there was a narrow headland here at Rackwick, but no sea stack. The cliffs were subsequently eroded by waves, and by 1817 there was a 'two-legged' Old Man of Hoy; we know this from an aquatint by William Daniell (1769-1837), an English landscape and marine painter. Daniell's painting shows a **wave-cut notch** at the base of the stack, and erosion here at sea level continued to undermine the structure.

Wave erosion won out, and sometime in the last hundred years, the seaward part of the Old Man collapsed, leaving only a single sea stack; the whole edifice is expected to collapse completely at any time, and nothing more than a rock stump will remain.

The beach at Rackwick Bay shows what happens to solid rock when it is subjected to marine erosion, and the foreshore comprises a range of sediments ranging from fine sand to very large boulders up to 1m (3.3ft) across. Most of this material has fallen from the cliffs as they were eroded, and the beach is the result of millennia of erosion and **weathering**. In time, the **cobbles** and **boulders** will be broken down further and carried away by waves and currents, to be deposited in other parts of the coastline, possibly hundreds of kilometres away.

The cliffs and beaches of Hoy are no sterile environment of rock and rubble, but a thriving habitat for plant and animal life, and the area here is designated a RSPB Nature Reserve. The Old Man is home to a flourishing colony of puffins, and it is arguably Orkney's prime site to watch what the locals call 'tammie norries'. Other seabirds that find a home in the rocks and crevasses include the northern fulmar, the great skua, red-throated divers, hen harriers and, most recently, golden and white-tailed eagles.

Geography and biology come together in the Orkneys. Puffins nest and breed on coastal cliffs and offshore islands. They build their nests in crevices among rocks or in burrows in the soil for protection against predators.

The beach at Rackwick Bay. Marine erosion and weathering have rounded these boulders, which highlights the sedimentary layers of Old Red Sandstone. The larger boulders weigh as much as 300kg (660lbs) and give an indication of the power of waves during storm conditions.

THE COASTAL ZONE

Before we go much further it will be useful to run through a few definitions that will feature in this book:

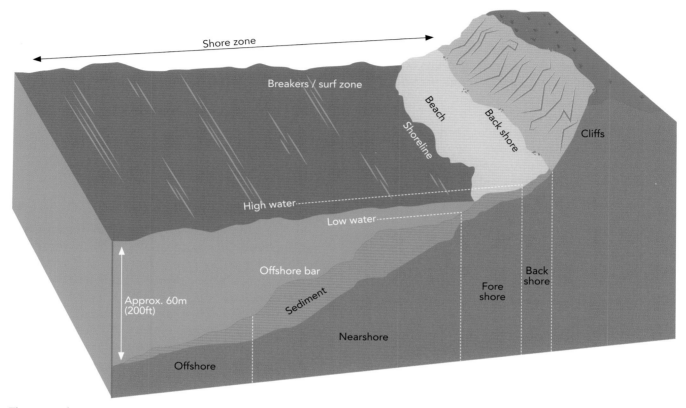

The coastal zone.

The **coastal zone** is the part of the land that is affected by its proximity to the sea, and that part of the sea that is affected by the land. This zone is subject to a wide variety of human activities, and the marine ecosystems here are some of the most vulnerable in our oceans, susceptible to marine pollution and irreversible damage.

The **shoreline** is the actual edge of the water – the boundary between land and sea. The shoreline therefore moves inland twice a day as the tide rises, and retreats as the tide falls. The shoreline also changes in the long term as some parts become eroded, and other places have stones, sand and mud deposited. In the very long term, sea level can change dramatically. The last glacial period (sometimes called the Ice Age) peaked around 21,000 years ago when most of northern Europe was covered in a vast layer of ice 3 to 4km (1.9 to

2.5 miles) thick. As the climate warmed, so the ice melted and the sea level rose by more than 120m (394ft), and this too has had a significant effect on the shape of our coastline.

The **shore zone** is the strip of coastline that extends from the top of the beach right out to sea to a water depth of around 60m (200ft); this is generally considered to be the depth at which ocean waves have no appreciable effect on the coastline. Here, the **offshore zone** is the most seaward part of the coastal zone, and this area of deep-water experiences little wave activity.

The **nearshore zone** extends offshore from the low-water mark. This area is affected by **longshore currents** as well as waves, making it a dynamic part of the shore where sediment is constantly on the move – inshore, offshore and along the shoreline.

The **foreshore** is the part of the beach that runs from low water to the top of the beach which gets wet from waves at high water; this limit can usually be identified by an upper line of seaweed. As the tides rise and fall, the waves affect different parts of the nearshore zone at different states of the tide. Wind speed and direction, and the angle and size of waves (especially during storm conditions) also shape the coastline in different ways.

The coastal zone presents a challenge for the survival of the plants and animals that inhabit the region. This thin strip of shoreline is under constant siege – not only from currents, wind and waves, but also from changing water salinity. Every six and a half hours, the tide floods the shoreline with salty water and bombards it with waves; the water then retreats, exposing the beach to the air, to the wind and to freshwater rain. The plant and animal life that live along the shoreline have to adapt to this ever-changing environment – and they have done so very successfully.

Waves have a very powerful effect on shaping our coastline. When fine-weather waves with a typical height of less than 1m (3.3ft) reach the beach, they dissipate an average of 10kW (ten one-bar electric heaters) of energy for every metre of coastline.

However, the energy in a wave is proportional to the square of its height, so a wave 3m high (10ft) has 3 x 3 = 9 times more energy than a 1m wave. During storms, the shore zone therefore experiences waves with the power to move rocks the size of a truck, and the force to demolish cliffs the size of cathedrals.

On a rocky foreshore there is constant erosion – this is nature's way of sandpapering the coastline which, in time, reduces the mightiest structure to powder. It is the product of this erosion, whether boulders, shingle, sand or mud, which is dumped around our coast to form the depositional features such as beaches, barrier islands and mudflats. These are discussed in the next chapter.

Rivers also bring dissolved and suspended material into the coastal zone and carry beneficial **nutrients** and harmful pollutants away from land and into the deep ocean. Vigorous water circulation from tidal currents and wave action mixes the coastal water with water in the deeper ocean. In this way, nutrients, pollutants and all the other material from the land (including plastic) eventually find their way into the most distant parts of the world's oceans.

A typical sandy beach at Yarmouth, Isle of Wight. The seaweed along the top of the beach marks the high point of the foreshore. Higher up are pebbles deposited by winter storms, and further up the backshore has been colonised and stabilised by coastal grasses.

WIND, WAVES, TIDES & CURRENTS

The British islands lie on the very north-western edge of the European continent and are therefore more exposed to the power of wind and waves coming in from the Atlantic Ocean than most other countries.

Oceanographers have produced what are essentially contour maps of average wave height and tidal height (amplitude) around our coastline. The prevailing winds create large waves which roll in from the Atlantic with extraordinary power to erode. In the north-western Scottish islands of the Outer Hebrides, for example, average wave height can be as much as 4m (13ft). Even Cornwall, south-western Wales and the Atlantic coastline of Ireland experience average waves heights of 3m (10ft) or more. Along the eastern coastline of the country, average wave is much smaller, typically less than 2m (6ft).

 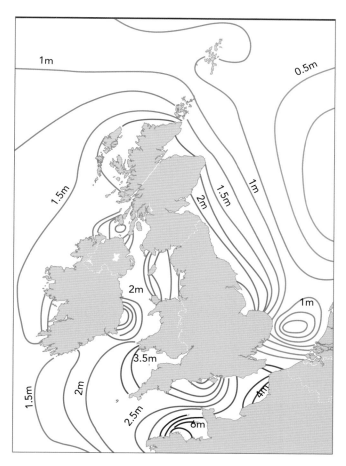

Average wave height (left), and tidal range (right) around Britain in metres.

Sediment is moved along a coastline by a combination of waves and currents. Waves lift sediment into suspension in the water, and currents carry the suspended material away. The map of tidal range gives some indication of the power of currents to move sediment from one part of the coastline to the other; generally, the higher the tide, the stronger the tidal current. Some of the strongest currents are found off the coast of Wales, the West Country and into the English Channel. However, in the North Sea, the tidal amplitude is typically 1.5m (5ft) or less, and the tidal currents (for the most part) are less. (Waves, currents and tides are looked at in more detail in chapter 4.)

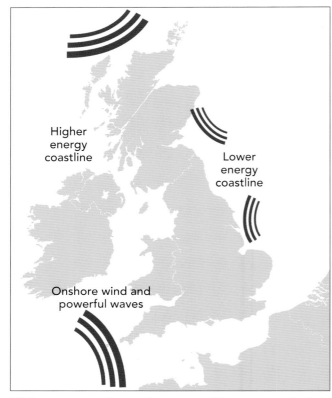

High energy coastline to the west, and lower energy in the east.

You can combine the wave height and tidal range maps to show high and low energy environments, and this creates two distinct coastal regions. Along the Atlantic seaboard from the West Country, north through Wales, Ireland and Scotland, the coastline is subject to powerful forces coming in from the North Atlantic. High energy waves, driven by the prevailing westerly winds, have the power to constantly shape

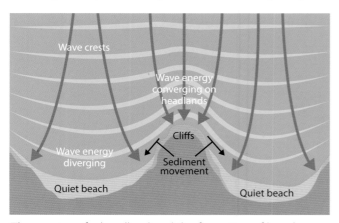

The erosion of a headland and the formation of beaches.

and re-form the shoreline; this creates a region predominantly sculpted by erosion. One cubic metre of water weighs as much as a small car, and this gives breaking waves enormous power to erode cliffs and move sediment.

Ceannabeinne beach in the Scottish Highlands; here rocky headlands are interspersed with sandy beaches.

Hunstanton beach in north Norfolk is an area of predominantly deposition.

Our coastal landscape is also determined by the type of underlying rock; the hardest and most resistant rocks are in the north and west of the British islands and, as you move towards the south-east, the rocks generally become younger and less resistant to erosion.

The oldest rocks in Britain are found in the north-west of Scotland, and some are nearly 3 billion years old – half the age of the Earth. The rocks of much of northern England and southern Wales are much younger limestones, which are moderately resistant to erosion. The underlying bedrock in southern and eastern England are mostly softer chalks, clays and sands, especially in East Anglia, and this makes the coastline more prone to erosion. As a general rule of thumb, the older a rock, the more resistant it is to erosion, although there are exceptions. (There is more about the geology of the British Isles and rock types in chapter 5.)

Over millions of years even the hardest of rocks will eventually succumb to erosion by waves, wind and rain. Even though the shoreline is often seen as a permanent fixture, changes can occur suddenly and dramatically, such as the collapse of the Old Man of Hoy. In some parts of the country – and in particular along the east coast of England – the young rocks can erode very quickly indeed.

Winterton-on-Sea in Norfolk, for example, is a popular holiday resort and many homeowners are keen to have a sea view which sometimes becomes too close for comfort. The underlying geology here is sands, gravels and silts laid down less than 4 million years ago – very recent in geological time. Because these deposits are largely unconsolidated, the sands are easily eroded and dozens of coastal houses along this stretch of the coastline have collapsed into the North Sea – a problem made even worse as sea levels continue to rise due to climate change.

Living on the edge at Winterton-on-Sea; in 2019, a report by the Global Commission on Adaption warned that thousands of British coastal homes will have to retreat inland.

CLIFFS & HEADLANDS

The most symbolic sea cliffs in the British Isles must be the White Cliffs of Dover. These stark, chalk cliffs rise from the English Channel and run inland to form a chalk ridge called the North Downs. To the south-west, a second chalk ridge starts at Beachy Head near Eastbourne and runs inland as the South Downs. This part of the south coast of England is a good example of how different types of rock respond to erosion and weathering, to produce very different coastal features.

Along the coast from Brighton to Eastbourne, the chalk ridge of the South Downs runs parallel with the coast of the English Channel. This fairly durable rock resists erosion to create a coastline with steep cliffs and prominent headlands, such as at Beachy Head. This is called a **concordant coastline**.

In the stretch of coastline between Eastbourne, and Folkestone the shoreline is very different. Alternating bands of 'hard' and 'soft' rock produce a discordant coastline, where wave activity and coastal currents nibble away at the more easily erodible clays and sands to produce bays and a low shoreline.

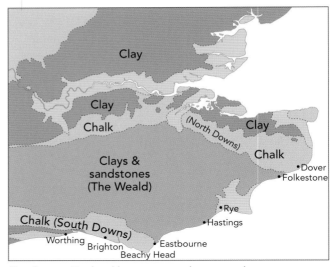

South-east England has a concordant coast between Worthing and Eastbourne where the chalk ridge of the South Downs runs parallel with the shore. To the north-east, a discordant coastline runs from Eastbourne to Folkestone, where clays and sandstones are less resistant to erosion.

Further up the Kent coast around Dover, the chalk of the North Downs is exposed, creating the White Cliffs of Dover.

The beach at Eastbourne is a wide, gently sloping beach of fine sand and mud formed from sediment carried along the coast by currents or brought down by rivers.

The White Cliffs of Dover rise to nearly 90m (300ft) above sea level. Chalk is a type of limestone made from billions of microscopic marine organisms laid down in a warm, shallow sea some 100-60 million years ago, when dinosaurs flourished on Earth.

We find cliffs all around our coastline wherever there is active wave erosion. As in Kent, soft rock, such as sandstone and clay, is eroded more easily to create gently sloping cliffs and bays. Harder rock, such as chalk, limestone, granite and basalt, is more resistant and erodes more slowly to create steeper cliffs. This process takes millions of years with hard rock, or just hundreds of years for softer rocks.

Cliffs are shaped and eroded by a variety of processes involving both **weathering** and **wave action**. **Corrasion** (also called **abrasion**) occurs when destructive waves hurl beach material, such as pebbles and sand, at the base of the cliff. This undermines the face of the cliff at sea level, and creates a **wave-cut notch**. This 'sandpaper effect' is particularly effective during stormy conditions.

Hydraulic action occurs when powerful waves strike a cliff and air is compressed into cracks; as the water recedes, the air pressure is released quickly, and this can cause cliff material to disintegrate. **Wedging** occurs when water in cracks freezes and expands, and so weakens the rock. **Corrosion** (rather than corrasion) is when specific types of rocks erode due to weak acids in water (especially rainwater); limestone (including chalk) is particularly prone to corrosion. Finally, **attrition** is when waves cause rocks and pebbles to rub together, causing them to break down into smaller particles; this is part of the process whereby cliffs are reduced to sand.

The processes of weathering and wave action work together very effectively to progressively weaken and erode the cliff face, which eventually collapses and retreats inland. The speed at which a cliff face recedes depends on

several factors, including the intensity of exposure to wave action. The hardest rocks, such as granite and basalt, erode slowly, whereas softer limestones and sandstones erode more quickly. The chalk cliffs along the south coast of England, for example, are eroding at a rate of up to 30cm (1ft) a year, and rising sea levels will accelerate this rate. By comparison, further west into Cornwall, the granite cliffs typically erode about 1mm (0.04in) a year, even though the West Country is more exposed to Atlantic storms.

As a cliff face disintegrates, two shoreline features remain. First, a flat, narrow shelf often forms at the base of the cliff from wave erosion. These **wave-cut platforms** are most clearly seen at low tide when they become visible as large areas of flat rock. Wave-cut platforms are usually covered at high water, so care should always be taken if you go inshore in a dinghy or kayak, as waves can steepen suddenly and sometimes break when they reach shallow water. The rock on wave-cut platforms is often eroded smooth and is slippery when wet – even more so when covered by seaweed. So always take extra care when walking over this terrain. The landward side of the wave-cut platform can become covered with sand to form a beach; when this occurs, the rocky platform can only be seen at low water, or when storms remove the sand.

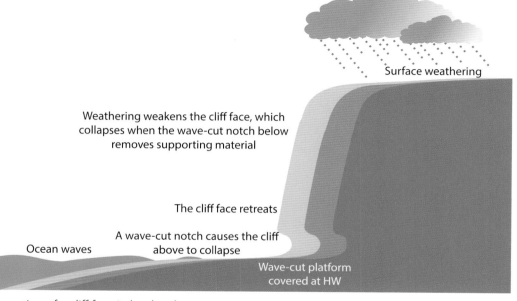

Surface weathering

Weathering weakens the cliff face, which collapses when the wave-cut notch below removes supporting material

The cliff face retreats

A wave-cut notch causes the cliff above to collapse

Ocean waves

Wave-cut platform covered at HW

The formation of a cliff face in hard rock.

Elgol beach on the Isle of Skye, showing a pronounced wave-cut notch. The rock face also shows characteristic honeycomb weathering of sandstone. On the beach, pebbles and large water-eroded boulders cover the wave-cut platform.

The other main feature from a retreating cliff is the rubble from the fallen cliff face. These boulders of varying size can most easily be seen on the beach above sea level, but this cliff debris also extends out to the sea, sometimes for several kilometres. This bottom terrain contributes to rough seas often experienced off headlands – an effect called **overfalls**. This is covered in chapter 4.

Cliffs are usually accompanied by other coastal features, including **caves, arches, stacks** and **stumps**. A headland is bombarded by waves from three directions; over time, it erodes and becomes smaller and, as it does so, other coastal features are formed. The famous Old Harry Rocks on the Isle of Purbeck in Dorset is a UNESCO World Heritage site, and the headland displays many of the coastal landforms associated with retreating sea cliffs.

Sea caves occur on almost every cliff headland, or along the coastline wherever waves break directly onto the cliff face. They are formed by **mechanical erosion** (hydraulic action), rather than through the chemical action that is responsible for most inland caves. Often waves of this size and power will carry shingle or even small boulders up into the cave, and these rocks can mechanically scour out the weakness in the cliff to enlarge a sea cave.

The foot of Beachy Head. The large grey rocks are weathered chalk boulders that have fallen from the cliff face. In time, these will be eroded into smaller particles and transported along the coast.

Labels on image:
Headland
Collapsed arch
Stack
Wave-cut notch
Wave-cut platform
Cave
Cliff face

The Old Harry Rocks on the Isle of Purbeck in Dorset shows a variety of coastal erosion features.

If a cave is formed in a narrow headland, it will eventually break through to form a **sea arch**, such as at Durdle Door in Dorset. Over time the arch will enlarge until it reaches the point when it can no longer support the top of the formation. When it collapses, it leaves a tall column of rock called a **sea stack**, which is isolated from the headland. A stack will continue to be eroded and eventually weakens and collapses to form a **stump**.

This is the process which created the sea stack at the Old Harry Rocks, and also the Old Man of Hoy.

The Dorset coastline is another example of a concordant coastline. Less than 2km (1.2 miles) east of Durdle Door is Lulworth Cove (also part of the Dorset and east Devon World Heritage Site). This is an excellent example of what happens when waves break through the hard rock strata (layers) of a concordant coastline. At Lulworth, the limestone ridge running along the Dorset coast has been breached, allowing incoming waves to erode the softer rocks inland. This is the best example of this feature anywhere in Europe and, for this reason alone, Lulworth Cove is visited by more school children and students than any other coastal locality in Britain.

At Durdle Door in Dorset, the narrow, limestone promontory has worn away to form a sea arch.

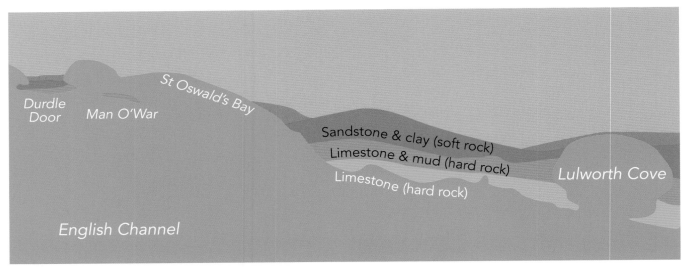

The limestone strata along the concordant coast of Dorset was breached, creating St Oswald's Bay and Lulworth Cove.

St Oswald's Bay looking south-east; remnants of the limestone ridge can be traced across the bay. The 'Man O'War' ridge (centre) and the distant stump beyond are all signs of active wave erosion.

CLIFF & ROCK SAFETY

Venturing along the coastline comes with a safety warning. Some sea cliffs, especially the softer sedimentary rocks, can be very unstable. The fossil-rich 'Jurassic coast' in Dorset is a very popular tourist destination, but the cliffs are liable to crumble at any time. In July 2017, 20,000 tonnes of cliff face collapsed onto the beach – fortunately without injury. The Coastguard warns that it is impossible to predict many of these cliff collapses and they urge beachgoers to be aware of the risks.

Cliff & Rock Safety

- Preferably go with a friend, check the weather forecast, and always let somebody know where you are going and when you expect to return.

- Check if the tide is coming in or going out, and also the time of high water; always be aware that your return could be cut off by a rising tide.

- Keep an eye on the waves – a breaking wave can easily sweep you off your feet.

- Carry a mobile phone – you can call the Coastguard on 999, but you cannot always rely on a good signal along the coastline.

- Carry water and wear appropriate boots and clothing; take sunscreen, a hat and sunglasses.

- Take extra care on slippery wet rocks. Barnacles and mussel shells can also inflict nasty cuts.

- Do not damage rock formations or hammer the stone; it is better and easier to simply forage among the beach boulders for interesting specimens.

- Do not touch strange objects and tell the Coastguard or police about them.

- Treat all cliffs as potentially dangerous; stay away from the edge at the top and also from the bottom of the cliff to avoid rockfalls.

- Do not climb cliffs unless you are an experienced climber with the correct equipment.

CHAPTER 2

BEACHES, BAYS & BARRIER ISLANDS

Marsden beach is one of many magnificent sandy beaches in the north-east of England. The sea stacks here are 255-million-year-old limestone. Souter lighthouse was the first to be designed to use AC electrical current, and in its day it was the most advanced lighthouse in the country.

A sandy beach on a sunny day is a huge playground, and it comes free to everyone. It gives endless pleasure to grown-ups and children alike, to surfers and swimmers, sailors and paddlers, fishermen, walkers and sunbathers. Beaches create memories too. We all recall idyllic days as a child messing around making sandcastles, and I am sure we also remember gritty sandwiches and wet towels full of sand. I suppose every silver lining has a cloud.

Pristine sandy beaches, however, are only part of the story. A 'beach' is really any foreshore where loose (unconsolidated) sediment has accumulted. The material can be made up from pretty much anything. On Skye, there is a brilliant white beach made from crushed algae, and at Seaham on the Durham coast the beach was once black from the spoil of coal mining. My own beach on the Isle of Wight is more eclectic – it comprises mud, sand, shingle, cobbles and even small fossils washed out from the clay cliffs.

The most obvious summer activities on beaches tend to obscure where the real action is happening – in the surf zone. Beaches are in perpetual motion: grain by grain, stone by stone, they are constantly changing their shape in response to wave and current action. Sandcastles collapse and footsteps disappear in no time. You need only spend an hour sitting on a sandy beach and you will see grains of sand blown by the wind, or rolled around in the surf; take a week, and you will see sandbars and wading pools created and disappear; spend a year, and dunes will grow then shrink; over decades and centuries, the beach landscape evolves on an even bigger scale.

Because beaches are continually being shaped by waves and currents, marine scientists usually consider that a 'beach' extends as far offshore to around 10m (33ft) below low water, for this is the normal limit of any effective wave action. Storms can make dramatic changes even further offshore, and in extreme cases they have been known to remove an entire beach of its sediment.

The mechanism that creates and changes beaches is straightforward. Solid particles – whether mud, sand, pebbles or larger cobbles – are moved by the turbulence of waves. If you get close to the edge of the water, you can actually see waves pick up sand particles and move them around. Because the sediment is suspended, it requires less energy to move than in air, so the material settles back more slowly. This allows heavier sediment to be carried further and for longer in water than in the air.

THE FORMATION OF A BEACH

If one section of coast is being eroded, then this sediment inevitably ends up being dumped elsewhere. This material is moved along the coast and deposited by waves and currents. When a wave approaches the shoreline, whether it is a headland or a beach, it breaks when it reaches shallow water. When this happens, the water in the surf zone physically moves the sediment forward towards the shoreline. (See chapter 4 for more on wave action.)

The water turbulence in the surf zone puts sediment into suspension, which can then be carried further along the shoreline by currents.

Each time a particle is lifted and settles in the wave zone, it does so in a slightly different place: it sometimes moves up the beach, sometimes down, or sometimes along the shoreline, depending on wave and current action. This constant shuffling and reshuffling of beach material, multiplied by the billions of grains of sand or pebbles on a beach, causes a beach to continuously change its shape and position.

Because waves are always moving beach material around, the shape of a beach depends on both the type of waves and the type of sediment. Mud and silt produce the flattest and smoothest beaches – typically no more than a slope of one or two degrees. This is because the finer mud particles are more evenly distributed when washed up on a beach, and the water then takes longer to percolate down through the fine deposits; this results in mud staying in suspension longer and being removed from the beach with the backwash.

As a general rule, sand produces wide, gently sloping beaches which are less than five degrees. Often, a sandy beach will slope gently at the water's edge, but the steepness increases up the beach as the sediment becomes coarser.

Beaches made from shingle and pebbles are steeper still, typically more than ten degrees. This is because the waves pass more easily through the porous, pebbly surface, and less sediment is carried back down the beach.

Breaking waves move substantial quantities of water towards a beach, called **swash**; as the water runs back down the beach, it is called **backwash**.

There is also a seasonal effect on the foreshore. Calmer conditions generally prevail during the summer, and these waves have a larger swash, and carry sediment far up the foreshore. These **constructive waves** build up a beach during the summer. During stormy conditions (which occur most often during winter), waves have a stronger backwash, and they remove beach sediment; these are called **destructive waves**.

Summer
Relatively gentle, low waves

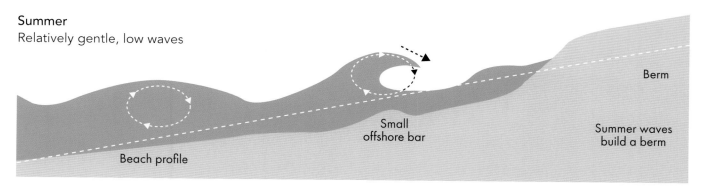

Berm

Small offshore bar

Summer waves build a berm

Beach profile

Winter
High, steep, powerful waves

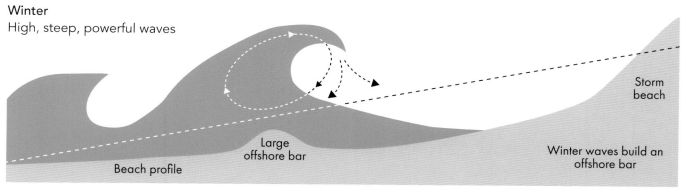

Storm beach

Large offshore bar

Winter waves build an offshore bar

Beach profile

In summer, gentler waves build a beach ridge, or berm; in winter, higher energy waves move material offshore.

Sandy beaches are not always smooth and often develop **ridges** and **runnels**. These are small sand waves found near low water which run parallel to the incoming waves and are formed through interaction between currents and waves in shallow water. The ridges are interrupted by little channels which allow water in the runnels to drain back down the beach.

These beach formations should not be confused with wind waves, which are created by the wind, and are found higher up the beach where the sand is drier and can be blown around.

 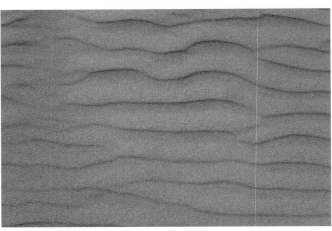

Ridges and runnels (left) run parallel to the beach and form close to the low-water mark; wind waves (right) are found higher up a beach, where the sand is drier, and the waves develop at right angles to the wind direction.

A rippled beach in Perranport, one of north Cornwall's most spectacular sandy beaches.

BEACH CUSPS

An intriguing shore formation are beach cusps, which usually (but not always) develop on beaches with coarse sands. The horns of the cusp are made of coarser material, and the embayment (shaped like a bay) contains finer sediment. They nearly always occur in a regular pattern, with cusps of roughly equal size and spacing. These features can be a few metres long or extend as much as 60m (197ft) across.

Marine scientists are not really sure how beach cusps form initially but, once they are established, they are usually self-sustaining as the waves interact with the shape of the cusp. As an incoming wave approaches a cusp, it splits and flows each side of the horn, the wave slows down and deposits coarser sediment on the horns. The waves then flow into the small bays carrying finer sediment, before crossing in the middle. After this collision, the water returns towards the sea as backwash, where it meets the next incoming wave. Therefore, once a cusp is established, coarser sediment is constantly being deposited on the horn, and finer sediment is being winnowed away from the embayment.

A complex cuspate beach at Bamburgh, Northumberland.

Coarse sediment deposited on the cusp horn

Backwash returns water back down the beach as a mini-rip current

Beach cusps are more common on beaches which face the open ocean, and therefore experience larger waves.

LONGSHORE CURRENTS

In the real world, waves rarely break exactly parallel to a beach, but do so at a slight angle. When this occurs, the breaking wave physically moves water along the shoreline, creating a **longshore current** – and with it, any sediment that is in suspension. When the current flows along a coastline with alternating headlands and bays, sediment stays suspended in the turbulent water off the headlands but is deposited in the quieter and less energetic conditions in a bay or estuary. The sediment carried along the shoreline is called **longshore drift**; it can be mud, sand and sometimes shingle.

A groyne on a pebble shoreline; these structures are designed to limit the movement of sediment along a beach. The beach is piled high against the groyne, indicating that sediment is moving from right to left.

Here the waves are breaking at a slight angle to the beach and sand is being moved along the beach towards the left. The groynes are trying to limit this.

The most obvious sign of longshore drift along our beaches is the presence of **groynes**, a low wall which extends down a beach and into the water, and made from timber, rocks or concrete. Groynes have some limited success at preventing the movement of sediment along the shore because they slow the movement of material along a beach. However, this frequently leads to more groynes or jetties being built further 'downstream' along the coast, to trap what little beach material might still be available. Sea walls also protect the shore from the relentless erosion of the coastline.

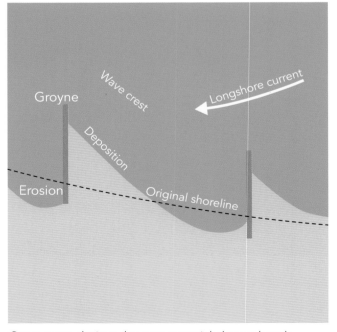

Groynes are designed to trap material along a beach.

COASTAL SPITS

The coastlines of Suffolk and Essex show several well-developed examples of deposition resulting from longshore drift. At Aldeburgh, the River Ore is diverted south by a huge shingle bank or barrier spit, which has built across the mouth of the river. The changing shape of the spit has been well documented for more than 400 years, and today the shingle ridge is between 3 and 6m (10 to 20ft) high. Inshore of the peninsula is an area of salt marsh which supports an internationally important nature reserve with rare and fragile vegetation, and a large bird sanctuary. This remote part of East Anglia was used for testing of early radar by the Ministry of Defence before the Second World War and continued to be used for sensitive military research during the Cold War.

This shingle bank, Orford Ness, is now owned by the National Trust and is open to the public, although access is strictly controlled to protect the fragile habitats. In protected coastal sites such as these, warning signs should always be respected.

The shingle spit at Orford Ness; over the centuries this feature has moved and grown. A ness is simply a headland and is derived from the Old English for 'nose'.

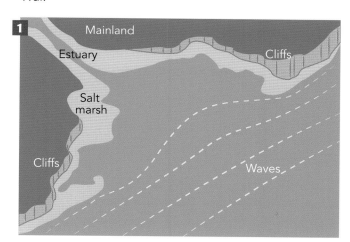

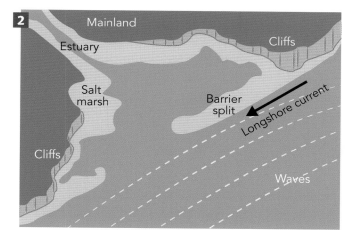

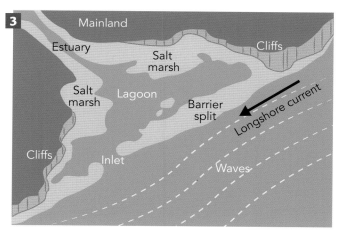

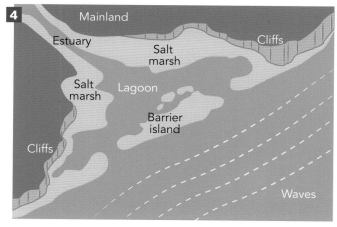

The formation of a barrier spit and island. As longshore current brings sediment along the coast, the material is deposited in the quiet conditions of a bay, when the speed of the current falls. Longshore drift is also responsible for creating a variety of low-lying features found along the British coastline, including barrier beaches, barrier islands, baymouth bars, and cuspate spits.

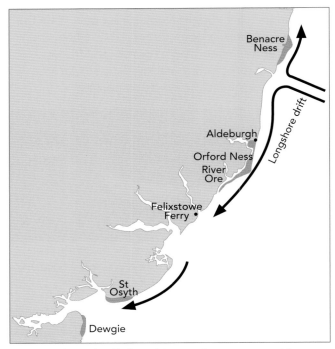

The coastline of Suffolk and Essex, showing coastal features that result from longshore drift.

Other spits have developed along the East Anglian coastline at Blakney Point and at Great Yarmouth in Norfolk, Felixstowe Ferry on the River Deben in Suffolk, and St Osyth in Essex. These coastal features often create family-friendly beaches and protected anchorages for sailors.

Spurn Head in Yorkshire is a **recurved spit**, where the end of the ridge curves around into the estuary. This usually occurs because the river flow out of the estuary is not powerful enough to wash away all of the sediment brought down the coast by the longshore current.

These coastal features are in a constant state of movement and are prone to being breached during storms, which happened at Spurn in 2013. A severe storm made the road to the end of the spit impassable to vehicles at high water and access to the RNLI lifeboat station there difficult at high tide.

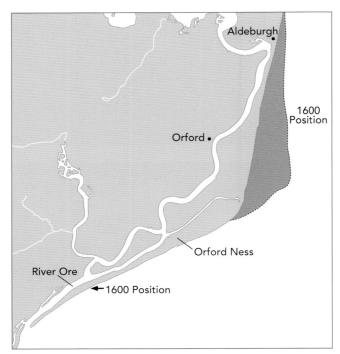

The changing shape of Orford Ness over more than 400 years. In the early seventeenth century, the ness extended much further out to sea, but less far south. Since then, the length of the ness has grown and the mouth of the River Ore has moved further down the coast.

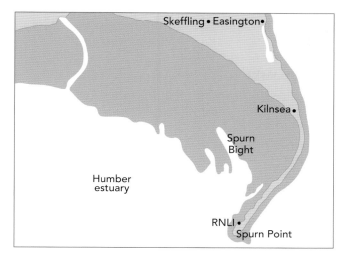

Spurn forms the northern bank of the mouth of the Humber estuary and is over 5km (3 miles) long, and sometimes only 46m (150ft) wide. The southernmost tip is known as Spurn Head or Spurn Point and is home to a RNLI lifeboat station and two disused lighthouses.

Further south in Essex, longshore currents have created a marshy region at Dengie, between the estuaries of the Blackwater and the Crouch. Here, currents have carried sand and mud into an area of low wave energy and a gently sloping beach. This feature is called a **chenier** or **chénier**, and typically forms a sandy or shelly beach, separated by mudflats; this creates an excellent marsh habitat for wildlife, and this nature reserve is now designated a Site of Special Scientific Interest (SSSI).

Under certain conditions, a sand or shingle spit will extend until it becomes attached to the mainland at both ends. This **barrier beach** then traps water behind in a **lagoon**: the best example in Britain is at Slapton in Devon. Others include: Bossington beach at Porlock, near Minehead in Somerset; Medmerry beach, near Selsey, West Sussex; and Oxwich Bay, on the Gower peninsula in Wales.

The 'low' light on the estuary side of Spurn Head dates from 1852. It was replaced by a taller lighthouse in 1895, which remained operational until 1985, when it was decommissioned.

Nairn beach in the Highlands, Scotland, looking south-west towards the inner Moray Firth. The spit shows clear signs of a westerly longshore drift of sand and a protected lagoon is beginning to develop behind the spit.

In many cases, these barrier beaches have (or at least once had) a lagoon on the landward side, and they often create a natural flood defence to low-lying land behind them. Like all low-lying coastal features, they can be breached easily, leaving them vulnerable to extensive flooding inland. It is obviously very serious if residential housing is flooded, but agricultural land too will become unusable for years if flooded with saltwater. Therefore, the conservation and protection of these low-lying coastal features is important.

Occasionally, a sand or shingle barrier beach extends from the mainland to an island, and this feature is called a **tombolo**.

A tombolo sand bar links St Agnes to Gugh in the Scilly Isles.

Slapton Sands is a barrier beach in Start Bay, south Devon. Inshore on the right is Slapton Ley, a brackish lake and nature reserve.

The biggest example of a tombolo in the British Isles is Chesil Beach. Here, a pebble strand stretches south-east from West Bay in Dorset to the Isle of Purbeck; the beach is 29km (18 miles) long, 200m (660ft) wide and 15m (50ft) high. For most of its length, the tombola is separated from the mainland by an area of shallow, brackish water called the Fleet Lagoon. Chesil Beach had a notorious reputation during the age of sail and has been the scene of many shipwrecks over the centuries. The beach was (and still is) particularly dangerous during south-westerly gales, where it forms a lee shore for ships heading east up the English Channel.

The fleet (Lagoon)

Sand

Shingle

Inceasing
sediment size

Chesil Beach

Prevailing
wind direction

Pebbles

Chesil Beach in Dorset, southern England, is one of three major shingle structures in Britain. Its name is derived from the Old English ceosel or cisel, meaning 'gravel' or 'shingle'. The prevailing south-westerly winds and longshore current carries along the coast, from west to east.

The pebbles on Chesil Beach are graded from pea-sized in the west and become progressively bigger towards the east.

On the southern Kent coast, almost equidistant between Hastings and Folkstone, is one of the most unusual stretches of coastline anywhere in the British Isles. The Dungeness headland is a **cuspate spit**; like many other depositional features, it is created by longshore currents bringing sediment along the coastline. However, a cuspate spit is formed when currents bring sediment from opposite directions. In the case of Dungeness, waves and their associated longshore currents deliver sediment along the coast from both the English Channel and from the southern North Sea.

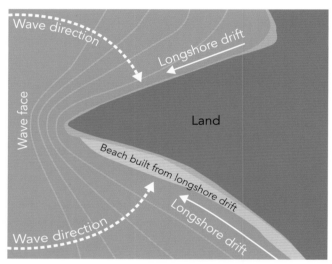

Dungeness in Kent; longshore drift brings sediment from opposite directions.

This unique part of the south coast of England is the largest expanse of shingle anywhere in Europe and has evolved over 10,000 years since the last Ice Age. After the ice melted, sea level rose quickly until about 6,000 years ago. By then, the area which is now Romney Marsh was a wide sandy bay; as the sea level rose, more sand was deposited by longshore currents, up to 10m (33ft) deep. The currents then deposited flint pebbles which had been eroded from the chalk cliffs of Dorset, Hampshire and Sussex during the Ice Age, building a barrier beach from Dungeness northwards to Dymchurch; behind this, the sandy bay became a brackish lagoon, rather like Chesil Beach today. Over the centuries, the lagoons filled with fine sediment brought down from the Kentish Weald, creating a salt marsh, with freshwater swamps in the valleys.

The Domesday Book of 1086 records that the whole of the main Romney Marsh area was occupied. The Marsh was still protected from the sea by a great shingle bank, but in the thirteenth century the sea breached the ridge, and the old town and port of Winchelsea was washed away. (It had been built on the shingle barrier somewhere around the current mouth of the River Rother.) In 1280, King Edward I ordered a new town to be built on 'the hill of Iham'; this is the town of Winchelsea we know today.

The sea continued to flood the area throughout the 1200s and 1300s, preventing any further settlement in the area; in 1349, the Black Death killed so many people that there was no longer any demand for new farming land for several generations. As the population grew again between 1400 and 1700, the salt marshes were gradually reclaimed for sheep pasture to supply the Kentish wool industry. By 1700, the area was similar to that of today.

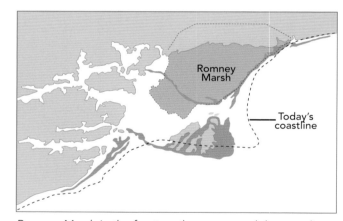

Romney Marsh in the fourteenth century, and the coastline today.

Dungeness now comprises a series of stony ridges and troughs, with the biggest pebbles found in the depressions and smaller pebbles on the crests. A wide range of plants have colonised the tops of these ridges because rainwater is trapped there by the smaller stones. This is also one of the best places in the country to find insects – moths, bees, beetles and spiders are attracted by the diverse vegetation. Many of these insects are rare species found nowhere else in Britain. The area also offers a refuge for many migratory and coastal birds, and the beach is renowned for excellent catches of winter cod.

Traditionally, the local community here were fishing people who launched their boats off the shingle beach in search of plaice and sole in the summer, mackerel in autumn, and herring and cod in the winter.

Among better known residents was the late film director, Derek Jarman, who grew both wild and cultivated plants in his pebble garden at Prospect Cottage. Jarman's tar-covered cottage is within a stone's-throw of the biggest building on Dungeness – a nuclear power station which first came online in 1965: Dungeness A has recently been de-commissioned but Dungeness B, a more advanced gas-cooled reactor, is still operational.

Derelict narrow-gauge tracks once brought wagons of fresh fish up from Dungeness beach to waiting trains.

Prospect Cottage on Dungeness, home of the late film maker, Derek Jarman.

The other buildings which dominate the Dungeness are lighthouses, and the headland has been home to no less than seven over the centuries. The first was built here in 1615 – a wooden tower with a coal brazier to provide the light. It lasted 20 years before it was pulled down and a new lighthouse built higher, and nearer the sea. As the pebble ness continued to grow, so the lighthouse became too far from the coast, and a third lighthouse was built in 1792. Since then, there have been another four lighthouses built – each one replacing a structure which became redundant as the ness continued to grow further out to sea. The current operational lighthouse was built in 1961.

The old 1904 Dungeness lighthouse with the 1961 lighthouse in the background.

Inland of Dungeness and protected by the shingle headland, is Romney Marsh; this is a good example of how a large coastal feature has been created by natural forces, and then tamed by enterprising farmers. The area is often called the 'Fifth Continent' and comprises nearly 260sq.km (100 square miles) of marshland, all of which is below the level of the highest tides. The area is protected from the sea by large earthen walls and natural shingle barriers. Although the area is very fertile, it relies on a complex drainage system to function as agricultural land.

OFFSHORE BARS & RIP CURRENTS

When waves approach a gently sloping shoreline, friction between the waves and the seabed causes the waves to break some distance from the beach. On these occasions, the breaking wave can deposit sediment offshore; as more material is deposited parallel to the coastline, a ridge called an **offshore bar** is formed. In these situations, waves can be seen breaking some distance off a beach before re-forming again as they come inshore. Offshore breaking waves are difficult to see from the seaward side, and care should always be taken when approaching these beaches in a dinghy, as the breaking waves can easily swamp a small boat.

As waves break along a beach, they feed a constant supply of water towards the shoreline. This water must return seaward eventually, but the position of an offshore bar can restrict this flow. In this situation, the surplus water can create a channel through the offshore bar, and a **rip current** is produced. (Rip currents are sometimes called 'rip tides' incorrectly, as they have nothing to do with the action of tides.)

Rip currents are relatively small-scale currents that appear in the surf zone, where water flows back out to sea from the shoreline. They can be permanent or temporary features, and their formation depends on wave height, frequency and direction. Permanent rip currents are found where a natural rock gully leads from the beach out to sea, whereas temporary rips are more difficult to predict. During a spell of heavy surf, rip currents can develop suddenly as they act as a temporary outlet for extra water building up on the beach front, but they then disappear once conditions return to normal. Rip currents can also migrate along the beach with a moving offshore sand bar.

Rip currents often result in the absence of surf, and the deeper water in the channel is often darker than the water on either side.

Waves are breaking on an offshore bar, and then re-forming before breaking again, but more gently, on the beach.

Rip currents are one of the most dangerous features on a beach for swimmers and divers and, in extreme cases, the current can reach 4 knots – faster even than an elite athlete can swim. Some beaches are more prone to rip currents than others, but these hazards have been recorded in almost every coastal location around the world. In the United Kingdom, over 60 percent of RNLI lifeguard incidents involve rip currents.

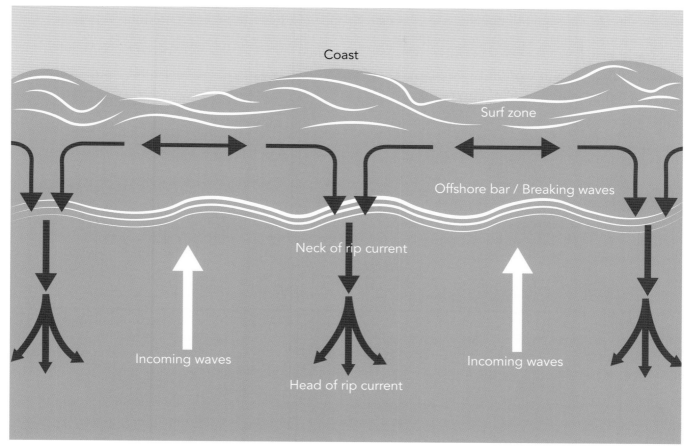

A rip current forms where surplus water brought towards a beach by breaking waves finds its way back out to sea through a channel in an offshore sand bar.

Rip Currents

Features of a rip current:

- The deep rip channel and the flow of water through it often results in the absence of surf.

- The feeder is the source of the water in a rip current, and even quite small points along the beach can make effective feeders; a rip current will typically have two feeders, although a single feeder is enough to sustain the feature.

- The neck is where water flows away from the beach, and can extend for several hundred metres offshore; it is usually between 10 and 30m (30 to 100ft) wide.

- The current in the neck flows away from the shore, making this the most dangerous part of a rip current.

- The head of a rip current is offshore of the bar. Here, it spreads outward and quickly dissipates. People who are swept this far offshore can find they have a long way to swim back to shore.

Swimmers who encounter rip currents often have little knowledge or experience in dealing with them. Because they do not understand how to respond, there is a risk that they panic as they are carried offshore. The initial reaction is to swim directly back to the beach, but this is rarely successful and results in the swimmer rapidly becoming exhausted.

Rip currents can occur on any day and at any location along our beaches, but they can become extremely dangerous under certain conditions. It is important to identify those days when rip currents are at their strongest and therefore the greatest threat to anyone who enters the surf.

How to Survive a Rip Current

- Do not overestimate your swimming abilities and stay out of the water when the surf is high.
- Learn to recognise the conditions which create rip currents; you can usually identify rip currents as rivers of clear water flowing away from the beach through the surf.
- If you are swept away from the beach, your chance of survival is improved if you stay composed and calmly assess the situation; do not try to swim back to shore fighting the rip; and try not to panic.
- Instead, swim parallel to the beach until you are out of the neck of the rip current. If this is not possible, then remain calmly afloat until the rip current dissipates, which it does at the 'head' of the rip channel, typically just beyond the breaking waves.
- Save your energy by ducking under the waves; breaking waves are powerful and if you duck under the wave, you will avoid its impact – this technique is used by surfers, and it can save your life.

Large, breaking waves move substantial quantities of water towards a beach (sometimes called the **swash**); as the water runs back down the beach, it is called the **backwash**. This movement of water back out to sea creates **undertow**, which can easily be felt underfoot, especially on steep, pebbly beaches. Bathers might be tumbled around in the water, but this return flow only lasts until the next breaking wave arrives. Undertow is experienced to a greater or lesser extent all along a beach, but unlike rip currents, it will not pull you offshore into deep water; undertow can, however, be dangerous for small children who might not be able to walk up the beach face against the strong flow of the backwash.

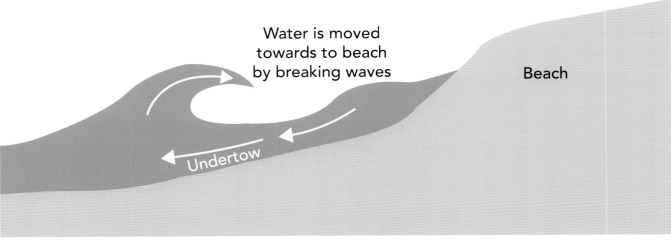

Undertow occurs where breaking waves carry water towards the shoreline, which inevitably results in a compensation flow of water back out to sea.

QUICKSAND

We have all seen movies where a hapless victim stumbles into quicksand. The more he struggles (and it always seems to be an ill-fated 'he'), then the more he is remorselessly sucked to his death – leaving nothing but his hat lying on the sand to mark his tragic demise.

Not surprisingly, reality is very different. In dry material, the 'quicksand effect' can certainly be fatal. In 2002, a farm worker in Germany fell into a grain store, and he was soon up to his armpits. As he breathed out, his chest volume decreased, and this caused the grains to fill the void, making it progressively harder for him to breathe. He soon experienced agonising chest pains. The victim was eventually freed, but only after his rescuers lowered a cylinder over his body to prevent the grains from re-forming around his body, and the grain was then sucked out with an industrial vacuum.

'Dry quicksand', so favoured by B-movie directors, is very different from the more common 'wet quicksand' that you find around the shores of the British Isles. Whenever we step ashore in a dinghy, or go wading onto mudflats, or even take a walk across a sandy beach, there is always the chance that we might plant a foot in quicksand. 'Wet quicksand' is a mixture of fine granular material (sand, silt, mud or clay) and water, and it can form wherever there is standing water, or when water flows upwards (such as from an underground spring).

The saturated sediment might appear quite smooth and solid on the surface, but any sudden change in pressure – such as when somebody puts their foot on the surface – causes the sediment to become liquified, thus losing its ability to support weight. The water effectively lubricates the grains, reduces friction, and causes it to change its characteristics. The cushioning effect of water also gives quicksand a spongy texture.

There is very little evidence that the more you struggle in wet quicksand, the further you sink. The reason is very simple – any objects in this liquified mixture will sink to the level at which the weight of the object equals the weight of the displaced sand and water mixture; at this point, the submerged object will 'float' due to its buoyancy. If the 'submerged object' happens to be a person, this is roughly about waist-level. So rule number one: do not panic, because you will not keep sinking.

Quicksand tends to become more viscous – in other words, thicker – the more you struggle. This happens because wet quicksand (which is initially quite liquid), becomes thicker as any movement squeezes out the water; the quicksand then begins to take on the characteristics of wet cement. Therefore, do not ask anyone to come to your aid in the quicksand, as they too could become stuck. The trick to extracting yourself is to wriggle to create a small gap around your legs; water will then flow into this space and loosen the sand. Take it easy, do not panic, and move your legs slowly and gradually.

The real danger in this situation is not the immediate effect of the quicksand, but the very real risk of becoming stuck with a rising tide and drowning. There are all too many cases where this has happened. In August 2016, two young men became trapped in quicksand on the beach at Camber Sands – a popular tourist area in East Sussex. Their three friends waded out to rescue them, but they too got into difficulties. Tragically all five men drowned in the rising tide.

Camber Sands in East Sussex. The very fine sand creates a beach with a slope of only a few degrees. Like many sandy beaches around the country, this idyllic looking strand harbours traps for the unwary.

Dealing with Quicksand

Your first defence against getting stuck, of course, is to avoid getting into trouble in the first place and to recognise where quicksand is likely to occur. If the worst happens, it is important to remember that the biggest risk is becoming stuck and drowning on a rising tide. So here are a few golden rules to avoid putting yourself in danger.

- If you have to walk across a suspect area, test the ground with a walking stick or branch; a few seconds prodding could make the difference between becoming a muddy mess, or having a safe trip home.
- As always when you find yourself in difficulty, try not to panic; you might sink, but you will not go deeper than waist level.
- If you are wearing a backpack or something heavy, take it off to lighten your weight; if you can, slip off your shoes or boots as it will be easier to get out.
- If you know ahead of time that you might encounter quicksand, take off your boots and walk barefoot.
- If you think you are becoming stuck, take a couple of gentle steps back before the quicksand takes hold. It usually takes a minute or so for the mixture to liquify and firm up.
- If your feet become stuck, move slowly – big steps only push your feet further down.
- If you cannot retrace your steps, then lean on your back and spread your weight – you will get muddy, but the alternative could be worse; once your body is supported, your feet should slowly come free, then roll sideways to bring your whole body to the surface.
- Somebody can pass you a rope or branch, but do not let your rescuer venture after you into the quicksand.
- You can also get out of wet quicksand by using your arms in a swimming motion, lying either on your front or your back (this keeps your mouth and nose clear).
- Be patient because you might not be able to extract yourself quickly; unless there is a secondary danger, such as a rising tide, nightfall or cold, there is no rush.
- It is worth stressing again that you should take your time and do not panic; desperate movements only make things worse and, whatever technique you use, do it slowly to stop yourself from becoming more deeply entrenched.

SAND DUNES

These are features along our coastline which constantly change shape, due mainly to the wind. They are formed when longshore drift moves sediment along the inter-tidal zone. At low water, the sand dries and onshore winds blow the sand up the beach. Dunes are most likely to occur where there is a large tidal range, and where the wind is predominantly onshore.

In time, an embryo dune will develop which may become covered with small grasses. Gradually, couch and marram grasses will become established. This vegetation stabilises the dunes in two ways: firstly, the roots bind the sand together, and secondly, the vegetation above ground traps particles of sand as they are blown over the surface.

Penhale Sands, north Cornwall; to the left are embryonic dunes, and inland are larger dunes stabilised by marram grass.

Marram grass beginning to establish in the dunes at Penhale Sands, north Cornwall.

Generally, the plants which thrive best in these conditions are those with succulent leaves which store water, with thorn-like leaves which reduce water loss in the strong winds, or those with long tap roots which can reach water at lower levels. Without the stabilisation effect of vegetation, dunes remain unstable and continue to migrate. Further inland, where the dunes are older, heather, gorse, reeds, rushes and other plants stabilise the sand even further. Eventually, oak, pine and birch trees take root, and this gradual development of heath and woodland inland is an ecological succession called a **psammosere**.

Sand dunes are fragile areas. Try to stay off them as much possible, as walking destabilises the dunes and destroys the plants. Elevated walkways can help

dunes from being trampled, and fencing will also trap the sand. On some beaches, owners drag their boats over the dunes, or store them on top of the dunes. This destroys vegetation, and when it has gone the area can become vulnerable to wind erosion. Likewise cutting or burning beach grass is very harmful.

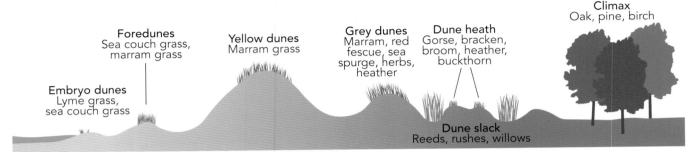

Embryo dunes
Lyme grass,
sea couch grass

Foredunes
Sea couch grass,
marram grass

Yellow dunes
Marram grass

Grey dunes
Marram, red
fescue, sea
spurge, herbs,
heather

Dune heath
Gorse, bracken,
broom, heather,
buckthorn

Climax
Oak, pine, birch

Dune slack
Reeds, rushes, willows

The formation of sand dunes on a beach.

A dune foreshore: marram grasses initially colonise the dunes, but small trees and bushes later establish on the older dunes. Behind is well-established woodland.

Like many depositional features around our coastline, sand dunes are a natural barrier to the power of wind and waves, and they form our first line of defence against coastal storms and coastal erosion. Dunes absorb the impact of storm surges and high waves, and they can prevent or delay flooding of inland areas. They also act as sand storage areas, and supply sand to eroded beaches during storms. We should never underestimate the value of sand dunes along our coastline.

BEACH SAFETY

Even relatively benign-looking beaches can catch out the unwary, and soft sands can trap walkers. It is therefore important to respect the coastline, understand the risks, and take appropriate precautions.

Beach Safety

- Preferably go with a friend, check the weather forecast, and always let somebody know where you are going and when you expect to return.
- Read warning signs carefully: a red beach warning flag could mean strong surf and currents, or it might mean that the beach is closed.
- Check if the tide is coming in or going out, and also the time of high water; always be aware that your return could be cut off by a rising tide.
- Keep an eye on the waves – a breaking wave can easily sweep you off your feet.
- Carry a mobile phone – you can call the Coastguard on 999, but you cannot always rely on a good signal along the coastline.
- Carry water and wear appropriate shoes and clothing; take sunscreen, a hat and sunglasses.
- Report missing lifesaving equipment, and anyone taking or breaking it.
- Do not leave litter on the beach: glass and opened cans are dangerous.
- Do not touch strange objects and tell the Coastguard or police about them.
- Sandy and muddy foreshores present their own risks, and you should always be aware of the telltale signs of quicksand.

CHAPTER 3
ESTUARIES, WETLANDS & SALT MARSHES

A fishing boat moored on the River Irt near Ravenglass, Cumbria. The estuary was an important Roman naval base in the second century called Glannaventa.

Eric Hiscock was one of the country's best known and most respected cruising sailors. Together with his wife Susan, the couple sailed around the world twice before retiring to New Zealand in the 1980s to live on their boat, *Wanderer V*. They wrote several books on small boat sailing and ocean cruising and gained immense respect for their sailing skills. Such was their extensive experience, it is difficult to believe they were once beginners.

Eric bought his first boat, an ageing 18-foot gaff cutter, in 1934 – long before he met his wife. His first cruise was from the Solent to the West Country. With no engine and little wind, it took him four days and five nights to sail to Brixham in Devon. He recalled one dramatic moment on the return journey, when he tried to enter the Salcombe estuary in a strong onshore wind:

> *The seas could not have been breaking properly on the bar, or I would not have survived; but one crest broke with sufficient force to fill Wanderer's large, open cockpit, and bursting open the cabin doors, flooded everything below. Fortunately, we were through the worst of it by then, for if another crest had come aboard, I think it would have been the end of us.*

Hiscock's recollection of entering Salcombe is a timely reminder that even when the sea appears to be benign there are always hazards whenever you are in shallow water, and it is always useful to be aware of what is under your boat, as well as what is above.

ESTUARIES

The estuaries around our shores, like Salcombe, are fascinating places: this is where rivers meet the sea, and where freshwater mixes with salt. Estuaries are nesting and feeding habitats for many insects, marine animals and birds. Most fish and shellfish around our shores spend at least part of their life cycles in estuaries, so they are crucial habitats for breeding and hatching.

Estuaries are also important commercial centres which have shaped our nation. London, Portsmouth, Southampton, Plymouth, Liverpool, Glasgow and Newcastle were all founded on big estuaries, and all these cities have made key contributions to the history and wealth of our country. Estuaries also offer a safe haven for ships during bad weather, and deep-water anchorages when it is time to disembark.

For marine scientists, estuaries are regions of transition between the land and the sea, where very distinct changes in temperature and salinity occur.

They are places where significant physical, chemical, biological and geological processes occur. Often, as much happens in just a few hours in an estuary that might occur in a hundred years in the deep ocean. As a result, estuaries are important, complex and dynamic parts of our coastline.

The entrance to Salcombe estuary, where Eric Hiscock almost ended his sailing days, looking much more peaceful.

Estuaries also offer a protected area to sail, swim, fish and walk. In some estuaries, such as the Camel in north Cornwall, the old railway track along the beautiful estuary has been turned into a walking and cycling path. Here you can pack a picnic and a beach towel and take off for the day, knowing the steep valley sides will give you more protection from the wind than you would ever get on a coastal beach. It is usually much easier to launch a dinghy or canoe in an estuary than on the coast, and the shallow, warm water is safer for little children to paddle and experience open water.

The Camel estuary, looking east towards the village of Rock.

The Camel Trail follows the course of an old railway line to Wadebridge and bicycles can be hired. (Taken from West Country Cruising Companion)

An estuary is usually defined by geographers as a partially-enclosed coastal body of water where seawater mixes with freshwater from the land. River water dilutes seawater, so salinity in an estuary varies between the ocean average of 35ppt (parts per thousand) and completely freshwater; the salinity will also vary according to the state of the tide.

Generally, estuaries are defined by their general shape, or topography, and you can identify two main types around the British Isles.

COASTAL PLAIN ESTUARIES

Estuaries are temporary marine features when viewed over a time scale of tens of thousands of years. For example, 18,000 years ago when the planet was in the grip of the last major Ice Age, the continents of both hemispheres lay under a thick layer of ice. So much water was locked up in these vast ice sheets that global sea level was up to 130m (430ft) lower than the present day (see page 12).

During this period, the land exposed around the margins of the ice sheet was eroded by rain and glacial meltwater, and rivers cut V-shaped valleys into the landscape. When global temperatures began to warm, the ice melted, and the continental ice sheets retreated. As a result, sea level rose, and these deeply incised river valleys were flooded, creating what is called a **coastal plain estuary** or drowned river valley. These are the most common estuaries around the south coast of the British Isles, and typical examples include the River Thames, Southampton Water and the Salcombe estuary.

In drowned river valleys (sometimes called **rias**), the deepest part is usually at the mouth and the estuary progressively becomes shallower and narrower inland. You sometimes find islands in these drowned river valleys, which are the summits of partly submerged hills which existed before the estuary was flooded. Coastal plain estuaries also tend to be relatively long compared to their width, although this also depends on the type of rock through which the original river valley was cut. The River Thames is the second longest river in England and 346km (215 miles) long, yet it is no more than 8km (5 miles) wide at its mouth at Sheerness.

The estuary of the River Erme in Devon, looking north toward Dartmoor. The steep-sided valley has filled in over the centuries with mud and sand. The South West Coast Path passes over the upper estuary, but crossing is only possible at low tide – and even then, you are likely to get wet knees.

Estuaries are generally places where sediment is deposited, and a layer of recent sediment of varying thickness usually covers the floor of the entire estuary. The type of sediment usually depends on the strength of the water flow in the estuary. You typically find shingle and coarser sands at the entrance where tides are stronger, becoming increasingly muddier (i.e. finer) in the upper reaches as the speed of water flow reduces.

One important feature of an estuary is the deep-water channel. Water flowing into and out of an estuary scours a deep channel on the outside of bends, whereas inside the bend, water velocities are reduced and sediment is deposited. This creates mud- and sandbanks, and a winding channel. When navigating an estuary or looking for somewhere to anchor for the night, the outside of bends will offer deeper water. However, always take care to anchor away from navigation channels.

Freshwater flow into an estuary can be substantial during heavy rain or spring thaw, but it is generally relatively insignificant compared to the volume of seawater which flows in and out with every tide. This can be used to advantage when entering an estuary. As a rule of thumb, it is best to enter in a boat on a

rising tide which offers a favourable current, and it will also lift you off the bottom should you inadvertently run aground in strange waters.

The Thames estuary from space, taken on 18 December 2003. You can trace the river's sediment far into the North Sea, carried on a northerly-flowing tide.

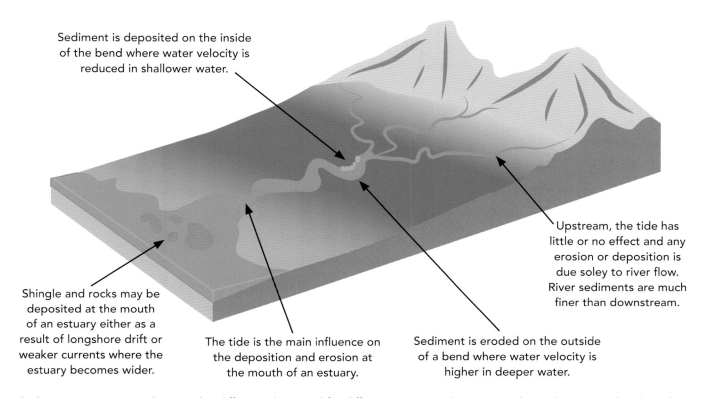

Sediment is deposited on the inside of the bend where water velocity is reduced in shallower water.

Shingle and rocks may be deposited at the mouth of an estuary either as a result of longshore drift or weaker currents where the estuary becomes wider.

The tide is the main influence on the deposition and erosion at the mouth of an estuary.

Sediment is eroded on the outside of a bend where water velocity is higher in deeper water.

Upstream, the tide has little or no effect and any erosion or deposition is due soley to river flow. River sediments are much finer than downstream.

Sediment in estuaries is deposited in different places and for different reasons. When water velocity decreases, shingle and gravel are the first to be deposited, and finer materials such as silt are the last.

Water is usually **stratified** or layered in an estuary because of the difference in density between fresh- and seawater. In summer, river water entering an estuary is warmer than seawater; being fresh, it is also less dense. Therefore, freshwater flowing into an estuary floats above the incoming, colder seawater. This makes a sandy beach in an estuary a great place for young children to paddle and learn to swim. This layering of the water in estuaries can result in a situation where the surface water can still be seen ebbing, whilst the incoming seawater below has already started to enter the estuary and water levels are rising.

These moored sailing boats are still pointing upstream into the surface ebb tide, but seawater could already be flowing into the estuary at depth on an early flood tide.

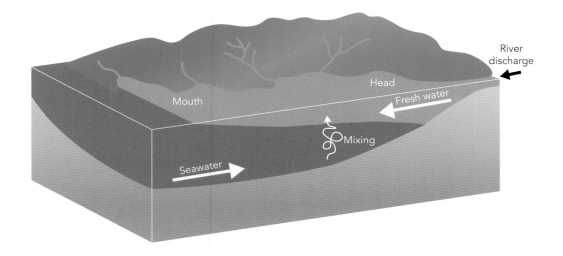

Less dense and warmer freshwater tends to float above the denser incoming seawater in an estuary.

The 'funnel' shape of coastal plain estuaries can affect the strength of the wind, especially when it is onshore. Wind always seeks the easiest route with the least friction; onshore winds tend to increase in strength at the entrance to estuaries due to the 'acceleration' effect from the topography, especially if there is high land on each side. The valley sides also cause the wind to funnel either up or down an estuary. As you move upstream, the wind is more likely to funnel through side valleys, therefore sailing up an estuary is often a series of 'spurts', interspersed with relatively calm patches.

Sometimes a drowned river valley is partially blocked by the movement of large quantities of sediment by longshore drift. This can form a characteristic bar across the entrance and creating a **bar-built estuary**. The River Deben on the east coast of England has a notorious shifting bar, where longshore drift brings sediment down the coast in a south-westerly direction (see chapter 2). The longshore current is interrupted by the tidal flow into and out of the estuary, and the sediment is dumped across the mouth of the river. Because the entrance is narrow, tidal currents can be high and, in the Deben, peak spring tides run at 4 to 5 knots, making it all but impossible to enter or leave the estuary against the tide in a small boat. Further upstream, where the tidal flow is greatly reduced, mudflats and sandbanks develop because of lower water velocities.

The River Deben, looking north. The shingle bar at low water is clearly visible.

Strong onshore winds and heavy breaking seas can move the Deben bar within a very short time. The entrance buoy and leading marks are moved annually, and sometimes during the summer sailing season in response to the ever-changing deep-water channel. Care must always be taken when entering this type of estuary as the navigation channels and markers may not always conform to the published chart. If in doubt, always take local advice. If this is not possible, a useful trick is to wait for another boat to enter the estuary (preferably one which has a deeper keel than your own). If the lead boat does not run aground, then you are also likely to be safe – but only if you follow their course precisely!

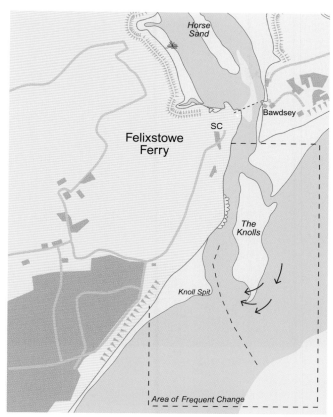

The River Deben entrance on the east coast of England; except for a narrow channel, the bar dries at low water. Longshore drift supplies the sediment that blocks the estuary.

Even if there is enough water to cross a bar, these underwater features are often associated with breaking seas. A submerged bar across the mouth of the Salcombe estuary nearly swamped Eric Hiscock's little sailing yacht, and these bars are potentially dangerous in a small boat and can make these estuaries a challenge to enter at low water, or with a strong onshore wind. Conditions are usually worst when a fast, ebbing tide meets a heavy onshore swell, combined with onshore winds. Even if the winds have died, a swell meeting an outgoing tide can still cause dangerous breaking waves.

This sounds like a depressing list of problems, so it should be remembered that thousands of boats every year visit the beautiful Salcombe and Deben estuaries. In settled conditions and with careful pilotage, there should be no problem to enter or leave any of these estuaries.

Sometimes, the streams and rivers flowing into bar-built estuaries have a very low water volume during most of the year. Under these conditions, the bars may grow into barrier beaches or barrier islands, and the estuary becomes permanently blocked. (For more information on barrier beaches and barrier islands, see chapter 2.)

GLACIAL VALLEYS (FJORDS)

Further north, where the land was covered by ice in the last Ice Age, glacial valleys were formed from the movement of ice (not water) and the valleys were eroded into a U-shape. The ice sheet came as far south as the Midlands, and this is why estuaries look very different in Scotland compared to England.

As the vast ice sheets covering Europe and North America slowly ground their way over the continental landmass, the ice eroded and widened pre-existing river valleys. When the ice cap and glaciers retreated, seawater flooded into the deep valleys, creating glacial valleys, or fjords. Because of the nature of their formation, fjords are found in high latitudes and in mountainous areas, and tend to be U-shaped in cross section, relatively narrow, and deep. In Norway – the land of fjords – they can run inland for as much as 100km (63 miles) or more. In the British Isles, most glacier-formed estuaries are found in Scotland, where they are called **sea lochs** (or loughs in Ireland).

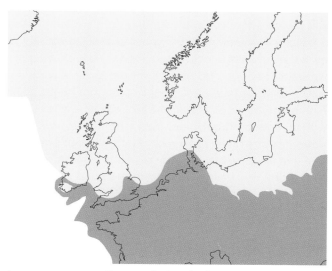

Ice coverage over Europe during the last Ice Age. The ice sheet was more than 1,000m (3,280ft) thick over Scotland, and sea level was 130m (425ft) below the level today.

The ice cover over Scotland looked much like Greenland today, with glaciers carving out U-shaped valleys.

Where a fjord joins the ocean, the entrance is usually much narrower than with drowned river valleys. A **sill** is often formed from terminal **moraine** (rocks and gravel) left over from the retreating glacier; sometimes this moraine is deposited on a ridge of solid rock. These sills or rocky bars are frequently found running right across the mouth of a fjord and also where a tributary fjord enters the main valley. The sills create a shallow entrance to a fjord and can cause problems for vessels entering or leaving the estuary, especially if a large ocean swell is running, causing waves to steepen and break over the shallow water.

Fjords and sea lochs often present a boat owner with an anchoring problem. Apart from being very deep, fjords usually have rocky bottoms with only a thin veneer of sediment. Any significant sediment deposition is usually restricted to the head of the fjord where the main river enters the valley, or where tributaries join the main valley.

River flow into fjords is usually small compared with the overall volume of water in the estuary. However, because the shallow sill restricts the tidal flow, river flow can often be significant compared to tidal flow, especially in late spring during ice melt. The rocky sill of a fjord limits the exchange of water between the estuary basin and the open sea. For the cruising sailor, the Scottish sea lochs offer spectacular scenery, but the deep water makes anchoring difficult. Even so, a shallow bay with a sandy bottom can usually be found.

Loch Hourn, sometimes described as the most fjord-like of the sea lochs in Scotland.

The largest glacial valley in the British Isles is Strangford Lough in Northern Ireland, which covers 150sq.km (58sq. miles). The name comes from the Old Norse *Strangr-fjörðr*, which means 'strong-inlet'.

WETLANDS & SALT MARSHES

As you progress up an estuary or venture behind sand and shingle spits, conditions change dramatically. Tidal flow and wave activity reduces, creating ideal conditions for **wetlands – tidal flats, salt marsh, peat bogs,** and **reed beds.** Tidal flats (also called mudflats) are coastal wetlands that form in the inter-tidal zone, and are created by the deposition of estuarine silts, clays and marine animal detritus. Being tidal, the flats are therefore submerged and exposed approximately twice daily, creating an unique environment for animals and plants. These salt-water ecological successions are called **haloseres**.

Tidal flats on the upper reaches of the River Deben near Woodbridge.

In the past, tidal flats were considered unhealthy places, fuelled by the belief that the 'the miasma of the marshes' brought disease and pestilence, and with good reason. The coastal marshes of eastern and southern England were perceived to be dangerous places, and deaths were unusually high in so-called 'marsh parishes', where 'marsh fever' and 'ague' (malaria) were endemic. This resulted in an epidemic between the sixteenth and nineteenth centuries, which resurfaced after the First and Second World Wars. The illness was probably caused by a northern strain of malaria, as well as tropical malaria. (The word malaria, incidently, comes from the Italian for 'bad air', *mal aria*.) Marsh parishes had such a bad reputation that only the poorest people would consider living there. One seventeenth century

geographer noted that Upchurch, on the Medway estuary, *'lies in the most unhealthy situation, close to the marshes'*, and in nearby Iwade, *'the stench of the mud in the ponds and ditches… contribute so much to its unwholesomeness, that almost everyone is terrifed to live in it.'*

With no economic value and a danger to public health, marshes were often drained to create agricultural land. Today, the tide of public opinion has changed, and the estuary flats and salt marshes are now considered to be important ecosystems supporting large numbers of marine flora and fauna; they are also a key habitat in allowing tens of millions of wading birds to migrate from breeding sites in the northern hemisphere to non-breeding areas in the southern hemisphere. Wetlands are also important in preventing coastal erosion, and help prevent flooding.

Salt marshes generally evolve from tidal flats. As sediment accumulates, so the flats grow in size and elevation. As the area becomes higher, so flooding is reduced, and this allows plants to colonise the area. Plants reduce the speed at which the river or creek flows into the sea, and this allows more sediment to settle. As sediment and plant species increases, so more sediment is retained, and over time new plant species become established.

Twice a day, a salt marsh is flooded by seawater at high tide, and then drains at low tide. This combination of regular flooding and fine sediment creates an environment with low oxygen, called **anaerobic conditions,** and promotes the growth of special bacteria. (This certainly added to the general air of menace in the marsh parishes.) The mud in salt marshes often has strong pungent odour similar to rotten eggs (derived from **hydrogen sulphide**), and this is a clear indication of low oxygen levels in the sediment. Another indicator is the colour of the mud, which is normally grey / brown. If you slice vertically through mud in a salt marsh with a spade and black layers are present, this indicates the iron in the sediment has been chemically altered to form iron sulphide, and these areas have little or no oxygen.

The Blackwater estuary, Essex; the tide runs more serenely inside the protected mouth of an estuary, and this allows extensive mudflats, salt marshes and reed beds to develop.

The regular flooding of salt marshes means the plants have to be very salt-tolerant, and species vary according to the degree of flooding. Plants growing on the lower reaches of the salt marsh – those parts which are covered by every incoming tide – need to be more salt-tolerant than those on the upper reaches, which might only be splashed by waves during high spring tides. This creates a clear zonation of plants, from the lower to the upper level of the salt marsh.

Mud is typically grey / brown colour; black is an indication of anaerobic conditions due to lack a of oxygen.

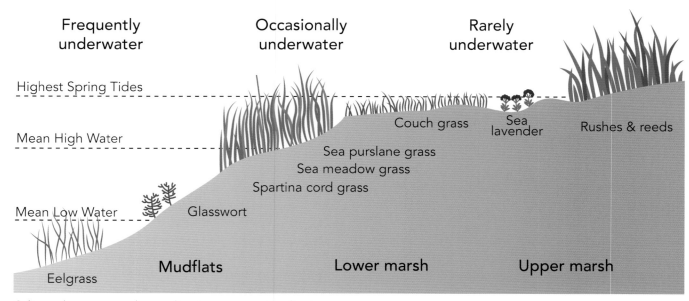

Salt marshes are zoned according to exposure to saltwater.

Microorganisms and tiny insects help break down the organic matter within salt marshes, and these then provide food for small fish, molluscs and crustaceans. During the flood tide, incoming ocean water brings nutrients into the area and takes plant material away during the ebb; this organic debris provides food for marine creatures offshore. On a rising tide, larger fish also move in and feed on the smaller species that permanently inhabit the salt marsh. Perhaps surprisingly, salt marshes are therefore one of the most diverse and productive ecosystems in the world. In them, rich organic growth is supported by nutrients coming down from the rivers, or flowing in from the sea. Organic production can reach 5 to 10 tonnes of organic matter per hectare (2.5 acres) a year, which is about the same as a field of wheat.

A salt marsh in Newtown Creek on the Isle of Wight at half-tide; the grasses are covered during high spring tides, and the creeks empty completely at low water.

The upper reaches of a salt marsh are colonised by reed beds, which are partically effective at trapping sediment and consolidating the salt marsh.

The Norfolk Broads are a unique area of wetlands in Britain, formed when rising sea level flooded Medieval peat excavations. Traditionally, the wetlands were drained by windmills (or, strictly, windpumps). The area is now a National Park and the country's largest protected wetland.

Salt marshes and wetlands offer a wonderful opportunity to explore unique environments and to watch birdlife, and there are lots of sites around the British coastline to visit. The Norfolk Broads is the biggest wetland in the British Isles, and it is easily accessible – you can even hire a boat for a family holiday. Cley Marshes, also in Norfolk, has often been described as a mecca for birdwatchers, and the extensive lagoons and reed beds are home to a huge range of birds. A new visitor centre overlooking the marsh, run by the Norfolk Wildlife Trust, provides great views. Minsmere on the Suffolk coast is the showpiece reserve for the Royal Society for the Protection of Birds (RSPB), and is famous for the 'Scrape' – a man-made lagoon with shingle islands, and home to nesting gulls, terns and avocets. Here you can also see other key wetland species such as the bittern and marsh harrier.

In Scotland, the Caerlaverock Wetland Centre on the northern side of the Solway Firth is one of the best places in Britain to watch wild geese and swans, which take advantage of the relatively mild winters and plenty of food. Common sightings include barnacle geese from Spitsbergen, and hundreds of whooper swans from Iceland.

The whooper swan (Cygnus cygnus) or common swan populations from Iceland regularly overwinter in salt marshes in Britain and Ireland.

Brancaster Staithe in north Norfolk; here the salt marsh is protected from the sea by the dunes in the far distance.

Wetlands Safety

These are fascinating places to explore and observe wildlife, but it is worth taking a few precautions:

- If you are alone, tell somebody where you are going and when you expect to return; always carry a phone, and remember the Coastguard can be called by dialling 999.

- As always, check the weather and be aware of a rising tide.

- Wear trousers and long sleeves as protection against spiky vegetation; insect repellent is also useful.

- Walking in marshy areas requires proper waterproof footwear which give good ankle support.

- Never walk without free hands, as you can easily lose your balance; a walking stick is useful to probe the area ahead and to check for quicksand.

- Wetlands are fragile ecosystems, so do not trample on sensitive areas and always respect nesting sites.

- Tidal areas are remote places and rarely visited, so you could always stumble on the unexpected; stay away from unusual-looking objects (which could be unexploded ordinance) and report the position to the authorities.

CHAPTER 4

WAVES, WHIRLPOOLS, SURGES & TIDES

Water in a breaking wave weighs around one tonne a cubic metre giving it enormous destructive power.

The morning of 31 January 1953 dawned unseasonably cold and cloudy, with squally showers sweeping across the country. A strong gale over Scotland veered to the north and increased to a severe storm. By the early evening, the east coast of England was taking the brunt of the tempest. A combination of these storm conditions combined with high spring tides and a storm surge raised sea level more than 2m (6.6ft) above the predicted height, causing severe destruction in low-lying areas bordering the southern North Sea.

WAVES

The storm of 1953 is a good example of how waves in the ocean come in many forms, and not just the breaking waves we see on the beach. There is the gentle slap of ripples against the hull of a boat, breaking waves so eagerly sought by surfers, giant waves in the Southern Ocean big enough to sink an oil tanker, and there are the monsters of the oceans – tsunamis – which can wipe out whole coastal villages. The tides too create a wave which floods our shores twice a day at high water. The way in which these different waves are formed might vary, but all waves are governed by the same physical laws of the ocean.

Waves that break around our coastline are created by wind blowing across the water. Wind transfers its energy to the surface of the water through friction; this causes small ripples to form on the surface and, if the wind continues to blow, then more energy is transferred, and the waves will grow.

Just how big any individual wind wave becomes depends on three main factors: the strength of the wind, the duration it blows, and the distance the wave travels – called the **fetch**.

The first waves to be created when wind begins to blow over a perfectly smooth water surface are **capillary waves** – or what we usually call ripples. **Surface tension** holds the water molecules together, which is the same force that allows insects to be supported on the surface of water. Capillary waves are very small, with a wavelength of less than 1.75cm (0.7in) between crests.

If wind speed increases, then the next stage of waves to develop are **gravity waves**; these overcome surface tension and become limited instead by the force of gravity. If the wind continues to blow, energy is continually transferred to the waves and the wave grows. If the wind dies away, the waves quickly disappear because of the combined effects of surface tension and gravity.

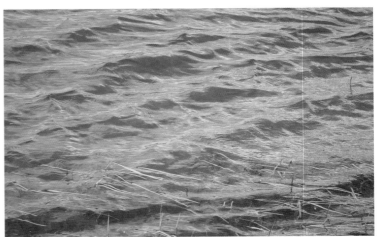

Capillary waves (top) begin to develop as soon as there is a light breeze but die away if the wind drops. If the wind continues to blow, capillary waves develop into gravity waves (bottom).

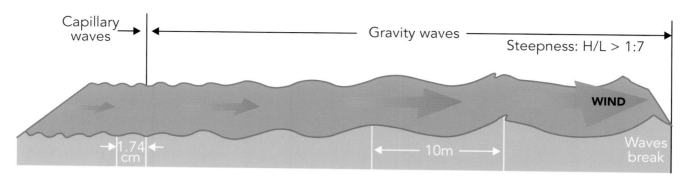

The development of a capillary wave to a breaking wave. Waves begin to break when the wave height to wavelength ratio exceeds 1:7.

Porthleven in south-west Cornwall takes a battering from Storm Imogen in February 2016. There is nothing between here and Miami except 6,840km (4,250 miles) of uninterrupted Atlantic Ocean.

Wave Terminology

Wavelength: the distance between two adjacent crests.

Crest: the highest point of a wave; **trough**: the lowest.

Wave height: the distance between trough and crest.

Wave frequency: the number of wave crests which pass in a given time period.

Wave period: the time that it takes a wave crest to move from point A to point B.

Amplitude: the maximum height of a wave from still water level; i.e. half the wave height.

Surface waves are called **progressive waves**, which are essentially any wave that can be seen moving through water. However, surface waves do not actually move water forward (at least not very much), but instead transfer energy – sometimes for very long distances. The water therefore becomes a medium through which **kinetic energy**, or 'energy in motion', passes.

The water in a wave is certainly moving, but only in a circular motion and not horizontally. This is why a gull sitting on the surface appears to bob up and down in the waves rather than move in the direction the waves appear to be travelling. Within the wave, the water particles are working like rollers on a conveyor belt – they rotate to move the belt forward, but the rollers themselves only rotate and do not move forward. (In practice, there is actually a very slight forward movement of water, but very little.)

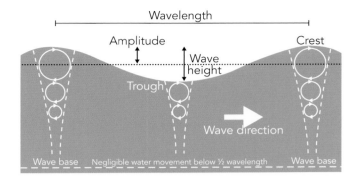

Anatomy of a progressive wave. Water within a wave moves in a circular motion.

A breaking wave is potentially very powerful – a cubic metre of seawater weighs over one tonne (one cubic yard weighs 1,728lbs) – by volume, this is as much a small truck. You become aware of the change in the motion of water particles when swimming off a beach. If you swim out into deep water and float, you will bob up and down as each deep-water wave passes, as it moves in a circular motion. However, once in shallow water in the surf zone, even quite modest breaking waves can knock you off your feet; this is because the water is physically moving forward in the wave.

We all experience how waves break as they 'touch bottom' every time we go to the beach. In deep water, the shape of a progressive wave is symmetrical for as long as the depth of water is at least twice the wavelength. Once the wave moves into water that is shallower than this, its characteristics change, and the wave begins to steepen. Wave steepness is the ratio of height to wavelength. When wave steepness is less than 1:7, breakers begin to form; when the ratio reaches 1:5, the wave become so steep that the crest topples, and the wave breaks fully. At this point, the water in the wave begins to move physically forward.

Waves generally add sediment to a beach during the summer (constructive waves) and erode a beach during the winter (destructive waves). Summer constructive waves have a lower wave height and a longer wavelength; therefore, the incoming water (swash) is dominant, and this builds up the beach (see page 33). Bigger waves with shorter wavelength occur in winter, and backwash becomes more dominant; this causes a beach to be eroded during the winter months.

Several factors ultimately limit the size of waves. When the speed of the wave equals the speed of the wind, then no more energy is transferred to the water, and the wave reaches its maximum size unless it interacts with something other than wind. If the

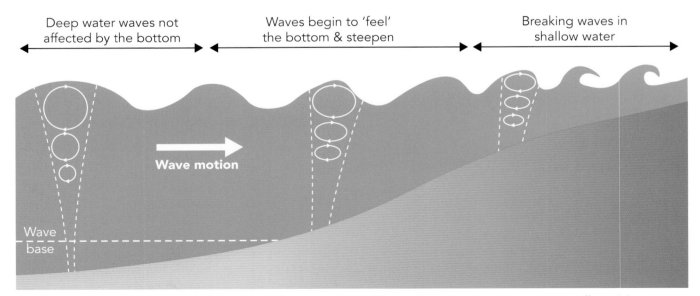

Breaking waves in shallow water. If the wavelength is more than half the water depth, then the wave is unaffected; however, as it moves into shallow water, the wave steepens. Once the wave steepness exceeds 1:5, the wave will break on the foreshore.

As a wave steepens, it becomes unstable and the top of the wave begins to break. At this point, water begins to move forwards towards the shoreline.

wind speed drops, then the wave will reduce in size. Waves will also decline when they become unstable in shallow water and break.

In practice, the seas around the British coastline comprise a mixture of many different sizes of waves, often moving in more than one direction. When a storm passes through an area, for example, waves are blown in different directions as the wind changes its track and intensity. This forms a **confused sea**.

White caps or **white horses** are a familiar sight in windy conditions. They form when the crest of a wave increases in size and steepness, becomes unstable, and collapses. Waves lose energy when their crests break, and this limits the size of the wave.

When waves move beyond the area where they are formed, they usually move faster than the local (gentler) winds; steepness then decreases. This results in a **swell**, which is a long-crested wave that has moved far from its source with little loss of (kinetic) energy. You do not have to go very far from the shoreline along the west coast, to experience swell coming in from the Atlantic.

A Royal Air Force Sea King helicopter attempts to lift an injured crew member from the deck of a French trawler in the Irish Sea in a force 8 gale. The boat is being raised 12m (40ft) with each swell. At this wind strength, foam is formed on the sea surface called spume, and the tops of waves are blown away as spindrift, reducing visibility.

Waves are also influenced by tidal currents, and the effect of **wind-over-tide** is familiar to anyone who goes to sea in small boats. If a current is moving in the **same** direction as the waves, it increases the wavelength relative to the height of the waves – effectively 'stretching out' the wave and making it flatter. The effect is a smoother sea – not because the height of the waves has reduced, but because the wavelength is longer, and the waves are therefore less steep.

Wind-against-tide: the wind is blowing from left to right in the same direction as the flood tide (left) and flattening the waves; four hours later (right), the tide has turned and the same wind strength is blowing against the ebb, causing choppy 'wind-against-tide' conditions and white horses.

BEAUFORT SCALE

Force	Speed kts	Speed mph	Breeze	Description
0	0-1	0-1	Calm	Sea like a mirror. Smoke rises vertically.
1	1-3	1-3	Light air	Ripples have a scaly appearance on water. Smoke drifts and flags indicate direction.
2	4-6	4-7	Light breeze	Small wavelets with glassy crests. Wind can be felt on the face. Flags indicate direction.
3	7-10	8-12	Gentle breeze	Large wavelets begin to break producing scattered white horses.
4	11-16	13-18	Moderate breeze	Small waves, becoming larger; frequent white horses.
5	17-21	19-24	Fresh winds	Moderate waves with regular white horses formed with spray. Flags fly horizontally.
6	22-27	25-31	Strong winds	Large waves with white foam crests and spray are extensive.
7	28-33	32-38	Near gale	Sea heaps up; white foam from breaking waves is blown in streaks.
8	34-40	39-46	Gale	Moderately high waves develop crests and begin to break into spindrift. The foam is blown in well-marked streaks along the direction of the wind.
9	41-47	47-54	Severe gale	High waves. Dense streaks of foam in the direction of the wind. Crests of waves begin to topple, tumble and roll over. Spray may affect visibility.
10	48-55	55-63	Storm	Very high waves with long over-hanging crests. Foam is blown in dense white streaks. The surface takes on a white appearance. Visibility affected.
11	56-63	64-72	Violent storm	Exceptionally high waves (small and medium-size ships might be lost to view behind the waves). The sea is completely covered with long patches of foam and the edges of wave crests are blown into froth. Visibility affected.
12	64-71	73-83	Hurricane	The air is filled with foam and spray. Sea completely white with driving spray; visibility very seriously affected.

ROGUE WAVES & TURBULENCE

In the jumbled-up conditions of the deep ocean, waves come from different directions and from different sources. Under these conditions it is inevitable that different wave patterns will mix and collide. This mixing of wave types creates an **interference pattern** in the ocean. These are not different types of waves, but a coalition of different wave patterns.

The interference pattern can be **constructive** where two or more wave patterns come into phase – this adds a crest to a crest, or a trough to a trough, to create a bigger wave. Waves also combine to become **destructive**, where a crest is added to a trough, making the wave smaller. Most commonly, a mixture where both constructive and destructive patterns are found together, and this explains why you might have a few large waves followed by a period of smaller waves.

Interference patterns can be unpredictable and extremely dangerous, creating what are commonly called **rogue waves**. These waves can be very large, but are usually short lived; although rare, they are by nature unpredictable, and capable of sweeping fishermen off rocks.

It is often claimed that every seventh wave is bigger than the others. However, the ocean is a chaotic system of many interacting forces, and it could never conform to such a rigid rule. Nevertheless, waves arriving at a beach on a fair-weather day will commonly arrive in groups of 12 to 16 waves. This pattern, coupled with the tendency for wave groups to 'bundle' their tallest waves in the centre of a wave train, provides a possible explanation for the seven-wave claim.

Turbulence in the water can make the sea surface rougher or smoother, depending on the conditions. This turbulence can be caused by water **upwelling** as it flows over a rough seabed, or around a headland. On a flat sea, the stirring action of turbulence will generate some waves, but they are usually small. If wind-driven waves move through this area, they may be destroyed by the turbulence and this can result in a distinctive oily-smooth disc of water.

Turbulence created by a rough seabed can create areas of upwelling. When this reaches the surface, it can stop wind-driven waves and create a spreading disc of oily-smooth water.

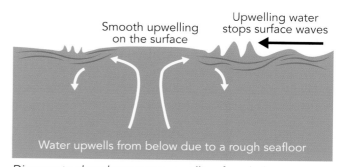

Diagram to show how water upwelling from below can create a smoothing effect on the surface.

Water turbulence off headlands can create a **tidal race** and related **overfalls**; these conditions are found when tidal currents increase in velocity off a headland, or sometimes in a narrow strait. The accelerated current flows over a shallow seabed which is usually uneven because of the eroded rocks left behind from the eroded headland. Shallow water, faster currents and a rough seabed combine to make turbulent sea conditions. If there is a strong wind blowing against the tide, then sea conditions will be even rougher, and nautical charts usually mark these areas clearly to warn inshore sailors of the hazard.

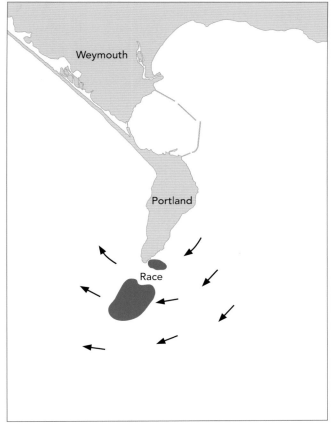

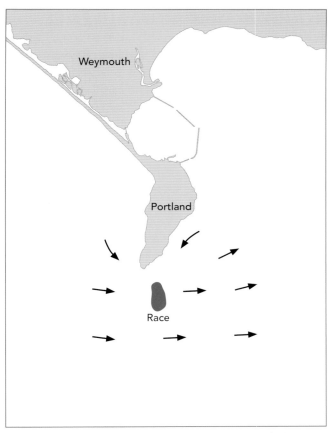

Portland Bill in Dorset has one of the most notorious tidal races and overfalls in the British Isles. The chart on the left shows a strong ebb tide. The ebb here can exceed 7 knots and, if it runs against a strong south-westerly breeze, sea conditions can become extremely dangerous. Six hours later the direction of the tide has changed (right) and the flood tide is running. Although not as strong as the ebb, tidal streams can still exceed 5 knots. It is worth noting that the position of the tidal race is not fixed but moves according to the direction and speed of the tidal current. At slack water, the race disappears completely, and it is usually best to round the Bill no more than a couple of hours either side of slack water, when conditions are at their most benign.

A fisherman's crane on Portland Bill, with the notorious overfalls in the distance. Here wind against tide combined with a rough seabed can create unpredictable and dangerous sea conditions, especially during spring tides.

WHIRLPOOLS & MAELSTROMS

In extreme cases where strong tidal currents accelerate through narrow, rough channels, a dramatic **whirlpool**, or **maelstrom**, can be created. These temporary features are at their most severe when the current is flowing strongly, and they disappear at slack water. The most famous whirlpool in the British Isles is the Corryvreckan, the third biggest in the world. (The name comes from the Gaelic, *Coire Bhreacain*, which means 'Cauldron of the Speckled Seas'.) The strong tide in the Gulf of Corryvreckan is caused by a two-hour tidal difference between the tides to the west in the open ocean, and the sound to the east; during spring tides the current can run up to 8.5 knots (about 10mph).

In 1947, George Orwell spent the summer on Jura to complete his masterpiece, *1984*. One afternoon, he took his son out for a boat trip in the Gulf, and the rough conditions in the maelstrom ripped the outboard motor off his dinghy, which then capsized. Father and son had to swim to a remote cove and wait to be rescued, and it was two hours before they were seen by a passing lobster fisherman.

Some years ago, a weighted dummy wearing a lifejacket and dive meter was lowered into the water and released close to Corryvreckan; it was eventually found several miles out to sea. The dive meter showed the manikin had been sucked to a depth of 200m (656ft) and gravel was found in the pockets of the torn lifejacket, suggesting it had been dragged along the seafloor.

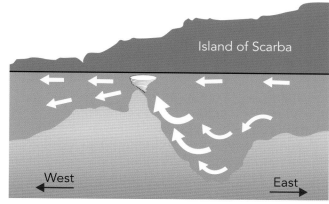

The floor of the Gulf of Corryvreckan rises from 219m (719ft) to just 30m (98ft) in a very short distance.

Safehaven Marine's Thunder Child II visits the Corryvreckan whirlpool during peak conditions.

STORM SURGES

The great North Sea flood of 1953 was caused by a combination of several factors, with different types of waves contributing to its overall destructive effect. The tide was particularly high in late January that year. In addition, there was also a storm surge; this is a mound of water created by low atmospheric pressure, which in 1953 raised sea level a further 50cm (20in) or so and was superimposed on the extra high gravitational tide. Finally, the strong northerly winds formed large, breaking waves more than 6m (20ft) high. The overall impact throughout the North Sea was devastating, and it became one of the worst peacetime disasters in Britain.

The impact of the surge was felt from Scotland, through eastern England, and into the continental low countries. In the North Channel between Stranraer to Larne, the ferry *Princess Victoria* foundered in the storm with the loss of 133 lives; only 44 people survived – none of them women or children. In England, 307 people died; 6,500 hectares (160,000 acres) of land were inundated with seawater and remained unusable for several years; 187,000 livestock animals were lost, and damage was estimated at £1.2 billion (in 2014 prices). The low lying coastal regions of Belgium and the Netherlands were hit even harder, and more than 1,800 people lost their lives in the Netherlands alone.

It was a timely reminder of the vulnerablity of coastal areas to the destructive power of tide and waves. The Netherlands embarked on extensive strengthing of coastal defences and surge barriers. In Britain, discussions began about how to defend London against a repeat event, although it took until 1982 before the Thames Barrier finally became fully operational.

The Thames Barrier became operational in 1982.

Blakney harbour on the north Norfolk coast. Raised sea level from tidal surges in the North Sea can cause havoc to boats lying alongside a jetty. The solution here is tall posts, which stop the boats from being washed onto the quay during a surge.

Strong winds

1.5m

1.5-2m

2m

2m

2.5m

1.5m

Water height above predicted sea level in metres in the southern North Sea during the storm surge of 1953.

TIDES

The other big influence on the 1953 storm surge were the extra high spring tides in late January. Tides are twice-daily events which affect every part of our coastline. They have a wavelength of up to half the circumference of the Earth, so they are the longest of all waves in the ocean; they are therefore considered to be shallow-water waves, and this influences how tides behave in coastal waters.

Because tides are created by the gravitational attraction of the Sun and Moon, they behave in a very predictable way, and modern computers can forecast tides for hundreds of years into the future. However, the relative position of the celestial bodies varies over the course of a month, and this gives rise to higher tides every two weeks called **spring tides**, alternating with lower tides called **neap tides**.

An understanding of the tides is important for anyone spending time around the British coastline. It will help someone in a boat get over a shallow bar safely, or prevent them from running aground; it will allow fishermen to avoid slack water when the fish are not feeding; and if you are rock-hopping along the shoreline it will let you know the best times to explore rock pools and not be cut off by a rising tide.

So, let us take the formation of the tides step by step and assume for the moment there is no Moon, and that we have only the Sun and the Earth to consider. (The Moon actually has a stronger influence on the tides, but the Sun allows a simpler explanation in the first instance.)

The Sun is 150 million km (93 million miles) away from Earth – so far that it takes light travelling at

Sailors mis-judge the tides at their peril.

300,000km/sec (186,000 miles/sec) eight minutes to reach us. However, the Sun also has a huge mass – about 330,000 times greater than the Earth – so it exerts a significant gravitation pull on our planet.

The Earth orbits the Sun once every 365.25 days, and that is why we have a leap year every four years to adjust for the quarter day. Our planet keeps to its orbit because the gravitational attraction of the Sun balances the tendency for the Earth to continue in a straight line out into the Solar System. This tendency to keep going in a straight line is called **inertia**, and for the Earth to maintain its orbit around the Sun, gravity and inertia must be in balance. This is the same principle that you experience on a playground carousel – we hold on to the merry-go-round (gravity) to keep us in 'orbit', otherwise we would spin off in a straight line (inertia).

The ocean, however, does not behave in the same way because water is both fluid and mobile. As the Earth orbits the Sun, the ocean on the side nearest the Sun is *very slightly* closer to the Sun than the centre of the Earth, so gravity is *very slightly* stronger; this extra gravitational pull causes the ocean to bulge towards the Sun. On the opposite side of the Earth, the water is *very slightly* further away from the Sun than the centre of the Earth, so inertia is *very slightly* stronger, and the ocean water creates a bulge as it tries to keep going in a straight line.

Over the course of a full 24-hour day, the Earth makes a single complete rotation about its axis. The ocean actually stays in the same position relative to the Sun,

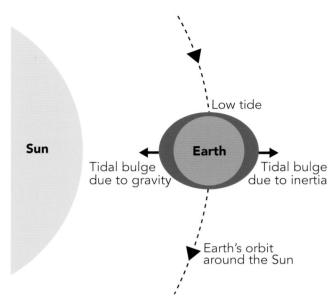

The Sun creates a slight bulge in the ocean water due to gravitational attraction, which is compensated by a second bulge on the opposite side of the Earth due to inertia.

and the Earth rotates underneath the ocean. This is not, however, how we perceive it. Standing on land and looking out at the ocean, we think we are static and the tidal wave (currents) appear to move past us. But, in reality, it is the ocean which remains static, and the Earth moves beneath the water.

If the tide is viewed from a fixed point on Earth, then sea level appears to rise and fall twice in a day. If you take the global tidal wave and straighten it out, it forms a wave with two periods of high water and two periods of low water every 24 hours, due to the Sun only.

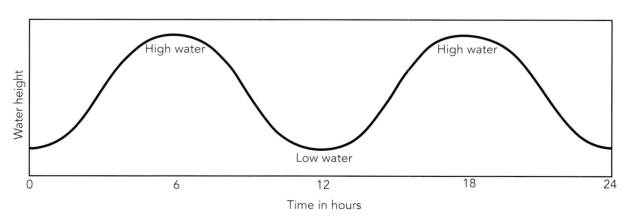

A simplified curve of a tidal wave; high water occurs twice a day as the Earth rotates in 24 hours.

In the real world, things get a little more complicated once you include the effect of the Moon. Although the mass of the Moon is tiny compared to the Sun, it is very much closer to Earth and has 44 percent more gravitational influence than the Sun. The Moon also orbits the Earth once every 27.3 days whereas the Earth rotates once every 24 hours, and these different time periods have a significant effect on the pattern of tides over the course of a month.

When the Sun and Moon are in alignment and their gravitational pull is in the same direction, the tides are at their highest; this situation creates a **spring tide** – this has nothing to do with the spring season, but the name comes from the tide 'springing forth'. When the Sun and Moon are out of synchronisation and their pull is at right-angles (which happens every seven days), this alignment produces a **neap tide**. The word 'neap' comes from the Old English word 'nep', meaning to become lower.

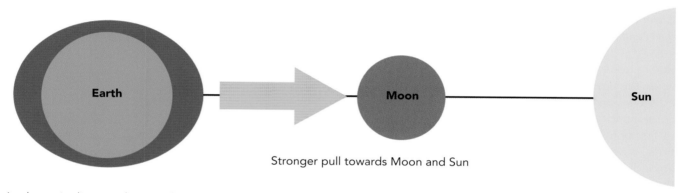

A schematic diagram showing the tidal bulges on the Earth's surface during spring tides, when the gravitational attraction of the Moon and Sun are in line.

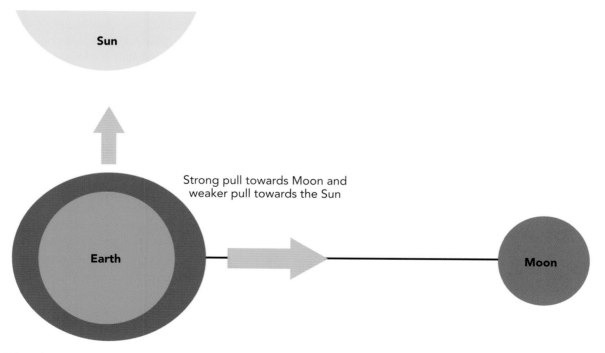

When the gravitational attraction of the Sun and Moon are at right angles, lower neap tides form.

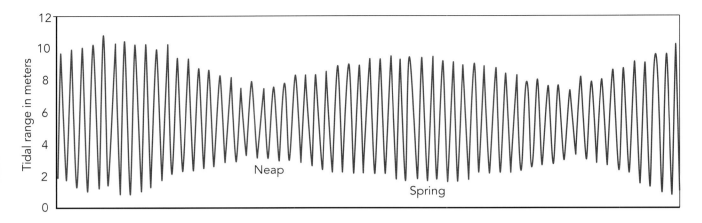

Example of tide data over a month showing alternating spring and neap tides. Here, spring tides are about 40 percent bigger than neap tides, which is fairly typical around the British Isles.

Spring tides therefore occur when there is a full Moon or no Moon, and neap tides correspond to the Moon in its first or third quarter. The tides, however, do not respond immediately to the passing of the Moon, so there is always a time lag of one to three days between, say, full Moon and the corresponding spring tide.

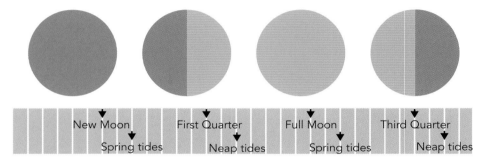

There is a time lag of 1 to 3 days between the phase of the Moon and its corresponding tide.

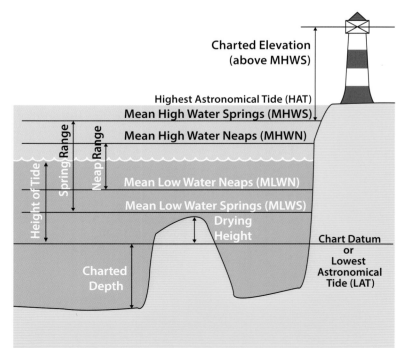

Because the Sun influences the tide on a 24-hour cycle and the Moon's cycle lasts 27.3 days, the times of high and low water are slightly longer than twice every day; in practice, the times between any two periods of high water occur about 12hrs 25mins apart, and this explains why the tides are about 50 minutes later every day.

There can be a large variation between high water and low water depending on location around the British Isles, and between spring and neap tides. This fluctuation is important to boat owners who must calculate the height of the tide to navigate safely. Marine charts use Lowest Astronomical Tide (LAT) as their base line; this is the lowest theoretical tide and lower than any anticipated spring tide. However atmospheric pressure can also change sea level by as much as 40cm (16in).

TIDES & SHALLOW WATER

A tidal wave conforms to the same laws of physics as a wave approaching shallow water on a beach. When a tidal wave moves into an estuary and the water becomes shallower, the shape of the tidal curve begins to change and steepen – just as a wave begins to steepen as it approaches a beach. As a result, the *shape* of a tidal wave will change as it moves further up an estuary.

The best example of this effect is in the Bristol Channel. As the tidal wave moves into the estuary, the frictional effect of the shallow water slows down the tidal curve on the flood, causing the tidal wave to steepen. At Avonmouth, the flood tide – i.e. the time from low and high water – is 5hr 23min, whereas the ebb tide lasts for 6hr 50min. Further up the estuary the difference is even greater, with the flood tide lasting 3hr 11min and the ebb tide 9hr 01min.

However, the same volume of water moves up the estuary during the flood tide that flows out on the ebb, but because the duration of the flood is significantly less than the ebb, there has to be a compensation. This causes the velocity of the flooding tide to be faster than the ebb. At Sharpness, tidal streams on the flood can exceed 8 knots (9.2mph or 14.8kph), or about twice the tidal stream on the ebb. (The ebb also has an element of river water, but this is insignificant compared to the volume of seawater being moved around.)

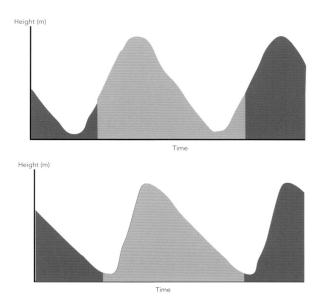

The same spring tide at two places on the Severn estuary. High water is about 40 minutes later further up the estuary.

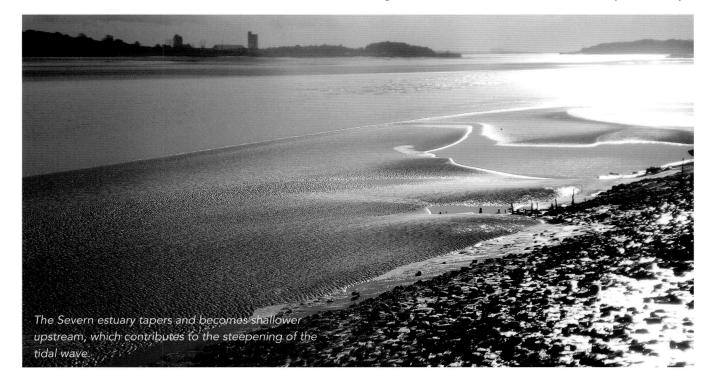

The Severn estuary tapers and becomes shallower upstream, which contributes to the steepening of the tidal wave.

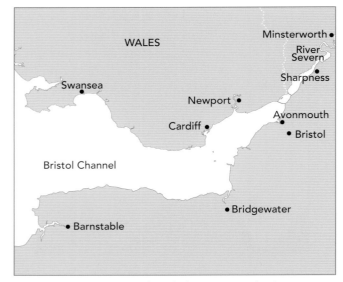

Map showing Avonmouth and Sharpness in the Severn estuary, relative to the Bristol Channel.

During high spring tides, the Severn estuary produces a **tidal bore**; this is a steep wave that is carried upstream by the rising tide. As a flooding tide enters the Severn estuary, the channel narrows significantly between Avonmouth and Chepstow; this, together with shallow water, causes the tidal wave to steepen. By the time it reaches the Severn estuary upstream of Chepstow, the width of the Bristol Channel has decreased from around 160km (100 miles) wide to less than 3km (about 2 miles). Here, the height of the tide can be as much as 15m (50ft) or more.

As the estuary continues to become narrower and shallower, so the bore becomes stronger. The bore consists of three or four quite sizeable waves, followed by smaller ones, and it is steep enough to surf; as with all waves, they will break in shallow water or near the bank.

There are about 260 Severn bores a year; the largest occur around the times of the spring and autumn equinoxes when tides are at their highest, but smaller ones can be seen throughout the year. Because the bores are linked to high spring tides, bores occur between 7am and noon, and between 7pm and midnight GMT, with the largest bores occurring between 9 and 11 in both morning and evening. The maximum bores occur a couple of days after new and full moons.

The bore can be seen from public footpaths along the riverbank, although there are often crowds at the more popular viewing places, and car parking can be a problem. A good viewpoint is at Minsterworth, at the appropriately named Severn Bore Inn on the A48.

As always, be aware of the power of breaking water and do not venture too close to the riverbank.

Tidal bores (sometimes called **aegirs**) can be seen in other estuaries around our coast, although none are as dramatic as the Severn bore. There are only around 60 of these phenomena seen around the world, and 11 occur in the United Kingdom.

Surfing the Severn bore.

The Severn bore will break in shallow water and along the river bank.

Tidal bores around the British Isles

- **River Dee**, England & Wales: the Dee bore can reach a height of 2m (6ft) and travels 2km (16 miles) upstream to Connar's Quay in Flintshire, Wales.

- **River Eden**, Cumbria, England: the bore travels 145km (90 miles) inland from the Solway Firth to the village of Wetheral in Cumbria, reaching 1m (3.3ft) and with speeds up to 14.5km/hr (9mph).

- **River Great Ouse**, Cambridgeshire/Norfolk, England: East Anglia's 230km (143 miles) river has a tidal bore locally called the Wiggenhall Wave; height up to 1m (3.3ft).

- **River Kent**, Cumbria, England: called the Arnside Bore, this short 32km (20 miles) river has a small tidal bore up to 30cm (1ft) high, and travels very slowly for 1.5km (1 mile) upstream.

- **River Lune,** Lancashire, England: small rolling waves up to 1m (3.3ft) high are seen between Lancaster through to the coastal village of Snatchems, 5km (3 miles) upstream.

- **River Mersey**, Merseyside/Cheshire, England: a substantial bore up to 5m (16.5ft) high and moving at 18km/hr (11mph) can be seen as far upstream as Warrington, a distance of about 48km (30 miles).

- **River Nith**, Dumfries & Galloway, Scotland: the only bore in Scotland reaches just under 1m (3.3ft), and moves 5.6km (3.5 miles) upstream from the Solway Firth to the village of Glencaple.

- **River Parret,** Dorset/Somerset, England: although the river is only 60km (37 miles) long, yet this tidal bore can reach heights of up to 2m (6.6ft), travelling at 10km/hr (6mph).

- **River Ribble**, Cumbria/Lancashire, England: slow rolling waves up to 1 m (3.3ft) can often be seen as far as Fishwick Bottoms, between Preston and Walton Le Dale, 18km (11 miles) upstream.

- **River Severn**, England & Wales: this is the world's second largest tidal bore and can be traced 40km (25 miles) upstream as far as Tewkesbury in Gloucestershire.

- **River Trent**, northern England: the bore can run for 80km (50 miles) as far as Gainsborough, traveling up to 19km/hr (12mph) and reaching a height of up to 2m (6.6ft).

The River Nith's tidal bore – somewhat less substantial than the Severn bore.

TIDES & THE WEATHER

The storm surge on 30 January 1953 and another significant surge on 5 December 2013, are timely reminders that weather can have a significant influence on the tides in several difference ways. The intense low pressure in 1953 raised the predicted sea level in the North Sea by around 50cm (20in). High pressure has the opposite effect and depresses the predicted level of the sea by about 1cm (0.4in) for every 1mb of pressure above the average of 1013mb; therefore, a fairly common high pressure of 1040mb will depress sea level by around 23cm (17.5in).

The wind too also affects the predicted height and time of the tide. Strong onshore winds increase the height of sea level and can also delay the time of high water; this effect is pronounced if the wind is blowing up a narrowing estuary. Strong offshore winds have the opposite effect by blowing water away from the coast and lowering sea level.

It is therefore always worth remembering that the published times of high and low water are only ever predicted times, based on a mathematical calculation of gravitational attraction. These timings will always be subject to weather conditions prevailing at the time.

Tides Around our Coastline

- The Venerable Bede (born around AD 673) realised that both the daily pattern of tides and the cycle of spring and neap tides were influenced by position of the Moon.

- The first tide predicting machine was designed in 1872; a later machine built in 1875-6 used 10 tidal components to predict the tides. Today's modern computers include about 62 constituents.

- The Moon's gravitational pull slows the Earth's rotation at a rate of 2.3 milliseconds each century, a phenomenon known as 'tidal braking'.

- Tidal forces also cause the Earth to distort by several centimetres, and big earthquakes are more likely to occur during full and new moons when tidal stresses are at their highest.

- Tidal power is the most predictable of all renewable energy sources and could generate up to 20 percent of the country's energy needs; eight sites have been identified around the coastline for barrages, but at the moment tidal energy is not considered to be financially viable.

- In the open ocean, tidal range is only about 0.5m (1.5ft), but the tidal wave steepens and increases as it approaches the coastline; the highest tides around the British Isles included the Severn estuary (up to 15m or 49ft), Morecambe Bay (10.5m or 34.5ft) and Dover (typically 6m or 20ft).

- The highest spring tide occurs once every 18.6 years and can create tides 0.5m (1.6ft) higher than normal; it last occurred in September 2015.

- Tide tables only offer a predication, but observed tides are influenced by many variable factors. For example, onshore winds can pile water up onto a coastline causing a higher tide than predicted, and offshore winds can lower sea level. High atmospheric pressure depresses sea levels, and low atmospheric pressure is associated with tides than are higher than predicted.

CHAPTER 5

COASTAL ROCKS & FOSSILS

Beachy Head in the afternoon sun clearly shows layers (strata) of chalk that were laid down between 100 and 64 million years ago.

In 1830, Scottish geologist and lawyer Charles Lyell, published *Principles of Geology* – a book which caused a seismic wave to shake the perceived wisdom about the formation and age of the Earth. Until this time, the theory of 'catastrophism' was popular: this suggested the Earth was created by sudden, short-lived and ferocious events, such as great floods and violent earthquakes. These cataclysmic geological episodes, it was claimed, created mountain ranges, and caused plants and animals to become suddenly extinct – to be replaced quickly by new species.

Lyell challenged these claims, and proposed that the world changed mostly in slow incremental stages such as erosion, and this created most of the Earth's geological features. He called his theory 'uniformitarianism', and suggested that the present was the key to the past – in other words, the Earth was shaped by the same natural processes that you see operating today. His theory fitted in tandem with Charles Darwin's emerging ideas about evolution.

COASTAL ROCKS

Since Lyell's day, sophisticated dating techniques and recent theories of seafloor spreading have given us a much clearer idea of how the Earth and continents were formed. We now know the Earth is around 4.6 billion years old, and during this time it has been bombarded by giant asteroids and subjected to volcanic eruptions and earthquakes. Sediments have been laid down in ancient oceans, only to be covered by more layers. These deposits were then buried deep in the Earth's interior, only to be brought to the surface by powerful mountain-building (tectonic) forces. To a geologist these ancient rocks are a picture book of the Earth's past, and the cliffs around the British coastline produce some of the most diverse outcrops anywhere in the world. The planet's geological history is on our doorstep.

The Headlands & Cliffs chapter showed how our coastal landscape is a response to the type of the underlying rock that is exposed, and the way it has been eroded. A map of the geology of British Isles shows a distinctive striped pattern, with the hardest and most resistant rocks in the north and west, with the oldest rocks in Britain found in north-west Scotland, some nearly 3 billion years old. As you move towards the south-east into England, the rocks generally become younger and less resistant to erosion. The rocks of much of northern England and southern Wales are limestones, which are moderately resistant to erosion, whereas the underlying bedrock in southern and eastern England are less-resistant chalks, clays and sands. These different rocks have a dramatic effect on the type of coastline we find around our shores.

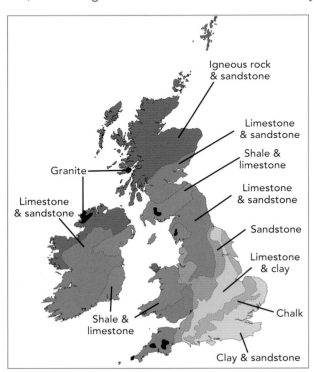

A simplified geological map of the British Isles, with harder rocks generally found to the north and west, and softer rocks towards the south and east.

The harder the rock, the more resistant it is to erosion, but over millions of years even the most robust rocks will eventually succumb to erosion by waves, wind and rain. The shoreline is often seen as a permanent fixture, but in some places where the bedrock is friable, change can be sudden and dramatic.

An afternoon's walk along a pebbly beach will reveal a wide range of different rocks. At first, trying to identify them might seem a daunting task, but in practice the basics are really quite straightforward, as there are only three types of rocks to be found anywhere in the world: **sedimentary**, **igneous**, and **metamorphic**.

SEDIMENTARY ROCKS

These are deposits that have settled at the bottom of a lake, shallow sea or ocean (and occasionally on the surface). They have subsequently been covered and compressed, sometimes for hundreds of millions of years. The sediments come from earlier rocks which were eroded to form **sandstones** and **clays**, or from the skeletons of minute marine creatures to create **limestones**.

The Old Red Sandstone in the Orkney Islands, for example, is a widespread sandstone laid down around 400 million years ago – long before dinosaurs appeared – and it is exposed on the Scottish coastline between Dundee and Aberdeen, around the Moray Firth, and throughout Caithness in the north-east. The Orkney and Shetland islands are also mostly Old Red Sandstone, where deposits are more than 4,000m (13,000ft) thick. You do not even have to visit the north of Scotland to see this attractive sandstone, because many buildings in Perth, Tayside, Stirling and Stonehaven were built from Old Red Sandstone; in England, the rock was used in the market hall in Ross-on-Wye, and to build the castles at Shrewsbury and Goodrich. The colour, incidentally, comes from iron oxide in the rock.

In Dorset, the sandstones around Lulworth Cove through to Swanage Bay include Lower Greensand, Gault Clay, and the tougher deposits of the Upper Greensand. These sandstones and clays are much younger than Old Red Sandstone and were deposited between 140 and 100 million years ago in a calm, deep ocean. The clays and sands of Dorset and

The Old Man of Hoy from the sea. These sedimentary rocks are 415 to 355 million years old and are continental in origin, rather than marine. The strata in the Old Red Sandstone is clearly visible.

the Old Red Sandstone of Scotland were both laid down horizontally and have remained this way ever since. This makes it easy to identify the strata on an exposed cliff face, especially after a fresh rockfall.

Sedimentary rocks, however, do not always remain horizontal. Tectonic movements inside the Earth, sometimes lasting hundreds of millions of years, can heat, cool, squeeze and fold rocks into almost any permutation, and the consequences can be seen all along our coastline as a kaleidoscope of rock formations. The dramatic chevron folding at Millook Haven in Cornwall is a spectacular indicator of what can happen when these powerful forces are brought to bear on sedimentary rocks. This outcrop comprises sandstones and grey shales (marine mud) deposited in an ancient sea around 350 to 300 million years ago, and the folding later occurred during a period of collision between continental plates.

A close-up of folded limestone.

425-million-year-old seabed of limestone, embedded with fossils.

Chevron folding at Millook Haven on the north coast of Cornwall.

Sedimentary rocks can also be biological in origin. Limestone is mostly made from calcium carbonate and is an accumulation of the skeletal fragments of plankton, coral and molluscs (and sometimes bones and teeth of marine animals). The best-known limestone in England is Portland Stone from Dorset, deposited around 150 million years ago. It makes an excellent building material and has been used in the construction of Buckingham Palace, the Tower of London, the Bank of England, the British Museum and the Palace of Westminster.

The gleaming white cliffs of Dover and the Needles on the Isle of Wight are formed from chalk (a specific type of softer limestone). Chalk is composed primarily of the shells of minute, single-celled, calcium carbonate creatures (plankton) which lived in warm, shallow seas between 100 and 60 million years ago, at the end of the dinosaur period. Chalk appears white because the skeletons of these marine plankton are colourless and reflect all light. When you stand back and look at a chalk cliff, you are actually looking at an ancient seabed made almost entirely of micro-fossils. If you are lucky, you might find bigger marine fossils, such as molluscs, starfish, echinoids, ammonites, sponges and even fish.

IGNEOUS ROCKS

The second group of rocks are igneous rocks, which originate as molten magma from deep inside the Earth. Igneous rocks contain randomly arranged interlocking crystals, making them relatively easy to identify, and are formed in one of two ways. When lava is erupted at the surface, the rocks are called **extrusive**; if they solidify whilst still below the surface, they are **intrusive**. This is an important distinction as the speed of cooling determines the size of crystals formed and the type of igneous rock.

When lava erupts on or close to the surface, it cools quickly and the crystals which form within the rock are therefore small. There are several types of extrusive igneous rock, but the most common is **basalt**, which is hard and dark. Because basalt is so resistant, the neck or plug of ancient volcanoes is often all that is left standing long after the rest of the volcano has eroded away. If the ancient volcano is surrounded by sea, then a dramatic circular island with steep cliffs is often all that remains. Good examples are Ailsa Craig in the Firth of Clyde, Bass Rock in the Firth of Forth, and the island of Rockall, 300km (187 miles) off the west coast of Scotland.

The Kilt Rock basalt cliffs on the Isle of Skye, Scotland.

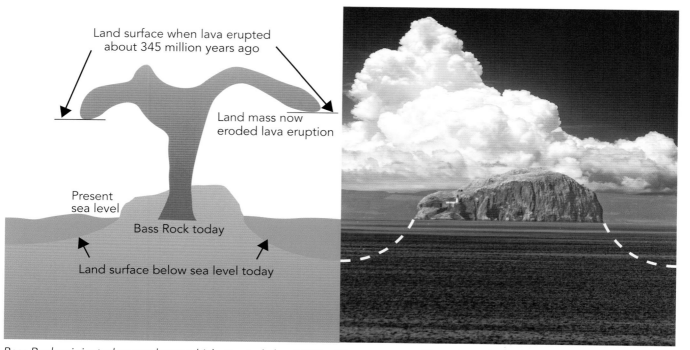

Land surface when lava erupted about 345 million years ago

Land mass now eroded lava eruption

Present sea level

Bass Rock today

Land surface below sea level today

Bass Rock originated as a volcano which erupted about 345 million years ago. Since then, the land surface has been eroded, leaving a volcanic plug rising out of the sea. The rock here is a rare extrusive igneous rock called phonolite.

The Giant's Causeway in County Antrim, Northern Ireland, is an area of around 40,000 basalt columns; most are hexagonal, although some have four, five, seven or eight sides. The unusual shape results from the contraction of the cooling lava which fractured, rather like the surface of drying mud; the cracks then propagated down as the lava cooled below the surface. This is now designated a UNESCO World Heritage Site

The basalt lava at Fingal's Cave on Staffa in the Inner Hebrides. The eruption occurred between 66 and 56 million years ago and formed hexagonally jointed basalt columns similar to the Giant's Causeway. You can visit by boat from Fionnphort on the Isle of Mull, from the Isle of Iona, or from Oban on the mainland.

If volcanic rock cools below the surface it does so slowly, and this gives more time for crystals to grow; therefore, intrusive igneous rocks have coarser crystals. The most common intrusive igneous rock is **granite**, although there are many others. We are all familiar with the large crystals in a granite kitchen worktop. Because igneous rocks are hard, they are very resistant to erosion, and form large coastal headlands and steep cliffs. Granite, for example, forms the spectacular coastline around Land's End in Cornwall, as well as the Isles of Scilly.

Granite cliffs in Cornwall.

METAMORPHIC ROCK

The third type of rock is metamorphic, formed from pre-existing deposits that are changed because of heat and /or pressure. Metamorphic rocks can originate from sedimentary, igneous or rocks which have previously been metamorphosed, and the crystalline structure of the rock changes as a result of being deeply buried or squeezed under immense pressure. During this process, the rocks become heated but do not melt; even so, the minerals within the rocks change, and this gives metamorphic rocks their unique crystalline structure. The oldest rock in Britain is a **gneiss**, found on the islands of north-west Scotland; it is nearly 3 billion years old, or two-thirds the age of the Earth.

Folded metamorphic gneiss on a coastal outcrop near Port of Ness on the Isle of Lewis, Scotland. These are the oldest rocks in Britain and are at least 2.7 billion years old.

Slate is another common metamorphic rock which is frequently seen in coastal outcrops. In Britain, the rock is typically between 400 and 300 million years old and formed by the low-grade metamorphism of mudstones laid down in deep water. Slate is the finest grained **foliated** (layered) metamorphic rock, which is why it is quarried as a high quality roofing tile. Slate has been an important extractive industry in Wales and Cornwall for centuries. The wreck of a sixteenth century ship in the Menai Straits was found to be carrying a cargo of slate, which indicates just how long the Welsh slate industry has been operational. There are also obvious advantages to quarrying slate on the coastline, as the cheapest form of transport in the past was always by sea.

The Blue Lagoon at Abereiddy in Wales was a working quarry until 1910, before being abandoned and flooded. The steeply dipping slate was laid down as a mud deposit about 465 million years ago. This is a good place to find fossils of the graptolite, Didymograptus murchisoni .

BEACH ROCK

All around our coastline are fabulous, easily accessible beaches which allow anybody with a little energy, an inquisitive nature, and an eye on the time of high tide, to explore. Here, you can learn a lot about the geology of the British Isles.

A beach can be made of a wide variety of sediments and sizes. In the Orkneys, the beach at Rackwick Bay comprises pebbles, cobbles, and even boulders more than 60cm (2ft) across; at North Landing at Flamborough Head in Yorkshire, a single step can take you from coarse sand to cobbles; and in Skye, the white Claigan Coral Beach is not actually made of coral, but from the dried nodules produced by a **coralline algae** called **maërl**.

North Landing at Flamborough Head, Yorkshire. The beach here ranges from coarse sand, through pebbles to large cobbles – all within a couple of paces. The beach is composed almost entirely of chalk pebbles, cobbles and boulders, together with some flint.

The brilliant white Claigan Coral Beach in the north-west of Skye.

The Claigan beach in close up, made from maërl. Unusually for a seaweed, maërl grows a hard, outer skeleton of lime which forms small-branched nodules, and it is these which are washed up to form a beach.

Black sand is commonly found in places such as Iceland and Hawaii, where active volcanoes are the source of the sediment. In the British Isles, the closest is probably at Talisker Bay – again on Skye – where there is a mix of black and white sand on the beach, often mottled to create patterns. The black sediment comes from ancient basalt lava which erupted about 60 million years ago. In other parts of the country a similar effect is produced from **magnetite**, a black mineral that is eroded from local rocks. Some dried algae can also create black streaks in sand, and of course there are also the dreaded tar deposits which periodically blight our beaches.

The distinctive black and white mottled beach at Talisker Bay in Skye.

Deposits which form a beach usually vary in size and type in different parts of a beach, and at different times of the year. Although it is not always the case, the smallest and lightest material is usually deposited near the water's edge, whereas bigger sediments – pebbles, cobbles, and boulders – are found higher up the foreshore.

The Wentworth scale used to classify and describe sediments by size

Type	Grain diameter	
	(mm)	(inches)
Boulder	≥256	≥10.0
Cobble	64 – 256	2.52 – 10.0
Pebble	4 – 64	0.157 – 2.52
Gravel (or Granule)	2 – 4	0.079 – 0.157
Sand	0.0625 – 2	0.0025 – 0.079
Silt	0.0039 – 0.0625	0.0002 – 0.0025
Clay	≤0.0039	≤0.0002

Rockfall from cliffs is fresh and tends to be sharp and angular, whereas pebbles and cobbles from the ocean are rounded from having been churned around in water for decades and centuries. Large, rounded beach rocks found high up the beach – such as those on Rackwick beach in the Orkneys – were deposited during a storm, when very big waves have the power to move these rocks weighing a third of a tonne or more.

A beach is the perfect place to go rock-spotting, and you can often find many different types of rock in a very small area. With a little knowledge and some basic equipment, you can identify different rock types. You need a hammer to split the stones, protective glasses, a magnifying glass and a pair of gloves. If you want to go into detail, a mineral guidebook or rock identification tables also help.

First, wet your rock, as you will then be able to see the minerals and structure more easily. If you have a hammer, split the pebble – it is essential to wear protective glasses, and also make sure nobody is standing close by. You will be able to see the sedimentary layers or minerals in a freshly-split pebble more clearly, making it easier to identify it as a sedimentary, igneous, or metamorphic rock.

SEDIMENTARY PEBBLES

Common sedimentary beach pebbles include **sandstone** which has hardened sandy or clay-like layers (strata), or **limestone** which is grey / brown whereas **chalk** is whitish. Most colours in sandstone come from iron staining – usually red, purple or yellow, but can sometimes be greenish:

- **Red** suggests the rock was deposited in a well-oxygenated environment such as in a river, flood plain or a shallow sea.
- **Green** means the rock was laid down in an environment low in oxygen – these are often deep-water marine environments.
- **Dark grey** to **black** indicates the rock was laid down in **anoxic** conditions (no oxygen), which might be deep water, but could also be a swampy environment.

You might find fossils, or water or wind marks in sedimentary rock.

A sandstone pebble with a light uniform texture.

Limestone pebble

Chalk pebble

Flint is found in chalk strata and is formed by the accumulation of quartz from the silica-rich skeletons of sponges and certain plankton.

IGNEOUS PEBBLES

Identifying igneous pebbles on the beach is relatively easy, as they contain mostly black, white or grey minerals, and have little texture or layering:

- If the crystals are visible to the naked eye, it is likely to be **granite** (intrusive), or possibly **gabbro** (usually black or dark green crystals).
- If the crystals are small, it is most likely **basalt** (extrusive), or occasionally **obsidian**, which is a black volcanic glass which has cooled very quickly within minimal crystal growth.

Igneous rocks cannot contain fossils.

The distinctive black and grey mottled minerals are easy to see in a granite pebble.

Much finer crystals are found in basalt; here water has exploited a weakness in this beach pebble to create a hole.

Gabbro is a coarse-grained intrusive igneous rock formed during slow cooling, and is chemically equivalent to the rapidly-cooled, fine-grained basalt.

Obsidian shows the distinctive glassy lustre typical of volcanic glass.

METAMORPHIC PEBBLES

Because metamorphic rocks have been changed by intense heat and pressure deep in the Earth's interior, the original rock is transformed into something much denser and more compact. This process creates new minerals which stand out in beach pebbles. Metamorphic rocks are often squashed, curved and folded into attractive layers of light and dark minerals. They come in a wide variety of colours, including red and green, and often contain glittery flakes of a silvery mineral called **mica**. Common metamorphic rocks include **phyllite**, **schist**, **gneiss**, **quartzite** and **marble** (a metamorphosed limestone).

Marble, showing the distinctive rounded shape of beach pebbles.

A metamorphic rock with distinctive flakes of shiny mica minerals.

A phyllite rock is usually light grey to blackish in colour. It has a slightly shiny lustre in the light, with slightly wrinkled or crinkled foliation planes.

Metamorphic rocks, such as schist, come in a variety of colours including blue, green and orange; schist displays attractive banding, typical of metamorphic rocks.

A selection of beach pebbles which can typically be found on many British beaches, including white limestone, iron-stained sandstone, mottled grey granite and multi-striped metamorphic rock. All of these pebbles originated in very different places, and have been transported long distances along the coast.

Quartzite pebbles are commonly found on the beach; this metamorphic rock is formed when quartz-rich sandstone or chert has been exposed to intense heat and pressure, which fuse the quartz grains together to create a dense, hard rock.

Beach pebbles are frequently found with holes in them, and these can result either from weathering and erosion, or from boring by a sea creature. If the pebble is igneous or metamorphic, then the hole is most likely to result from a weakness in the rock itself, as any flaw is vulnerable to erosion by the sea. Holes in softer, sedimentary rocks, are usually created by small invertebrate animals, including some species of bi-valve molluscs (such as **piddocks**), **polychaete** (bristle) worms, and even **sponges**. These are common along the Jurassic Coast in Dorset and in many other locations around the British Isles.

UNNATURAL 'PEBBLES'

Asphalt. Although not actually a rock, asphalt can also be very common on the beach.

Not all beach sediments have a natural origin. Even quite a short time looking carefully along most British beaches will produce a handful of **sea glass**. This is any glass that has found its way into the sea, where it has been physically tumbled by wave action and chemically weathered in seawater to produce beautiful, naturally frosted glass beads. Winter months are best to find these gems, when the beach is refreshed from rougher seas. Sea glass takes at least 20 years to acquire its shape and character, and often much longer; it can be collected to make jewellery or home decorations.

One of the best beaches in the British Isles to find sea glass is Seaham on the Durham coast, just 10km (6 miles) south of Sunderland. The town was home to the Londonderry Bottleworks (the largest glass factory in the country) and between 1853 and 1921 the company produced up to 20,000 hand-made bottles a week. At the end of the working day, unwanted or faulty glass was tossed over the cliffs into the sea. A hundred years on, gems of what was locally called 'end of day' glass are washed up on the beach every day.

Sea Glass

You can tell a lot about the origin of sea glass from its colour:

- **White** or **opaque** is the most common and accounts for about two-thirds of all sea glass; it could come from any source of clear glass, such as milk bottles, soda bottles or jam jars.

- Roughly a fifth of sea glass is **green** and it comes mainly from wine and beer bottles.

- **Brown** sea glass is also common and comes from a similar source, as well as old medicine bottles.

- **Purple** and **black** glass derives its colour from nickel, and **blue** comes from cobalt.

- Pre-First World War clear glass was made with manganese and, after many years, it turns **lavender** from being exposed to sunlight; this is one of the rarest colours to be found in sea glass.

Beach glass can be found in a variety of colours.

COASTAL FOSSILS

In 1811, a 12-year-old girl called Mary Anning was scouring the beach in Lyme Regis in Dorset with her older brother. The two young children often went fossil hunting to help supplement their meagre family income. That day they made an extraordinary discovery – partly exposed in a rockfall was the huge skull of a creature with big teeth. No one was really sure what it was because, at that time, people understood little about fossils, or even the history of the Earth. The skull became one of the first fossils to ever be described by geologists, and it was named an **ichthyosaur**, which means 'fish lizard'. Mary Anning sold the skull to a local collector for £23, or nearly £2,000 in today's money – an invaluable contribution to her family's financial reserves.

Mary went to make several more famous finds from the fossil-rich Jurassic beds of Lyme Regis. Although she was never trained, she gained both notoriety and respect within the scientific community and in 2010 – 163 years after her death – the Royal Society listed her as one of the ten British women who have most influenced the history of science. The 2020 feature film *Ammonite*, starring Kate Winslet as Mary Anning, is a dramatised account of her remarkable life, and her daily struggle of living on the poverty line. It has also been claimed that Mary Anning's story was the inspiration for the 1908 tongue-twister, 'She sells seashells on the seashore'.

Mary Anning died in 1847, aged 47. Her grave in Lyme Regis is a pilgrimage for dedicated fossil hunters, who leave their finds at the base of her headstone.

Mary Anning with the Golden Cap outcrop in the background. She very nearly died in 1833 in a landslide that killed her dog, Tray.

The Jurassic cliffs near West Bay in Dorset are famously fossil-rich, but care should always be taken as these vertical faces are prone to rockfalls.

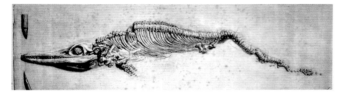

Drawing of an ichthyosaur skeleton similar to the fossil found by Mary Anning

You would be extremely lucky to find a complete ichthyosaur skull, but spectacular fossils are frequently found along the coastline of Britain. In 2016, a self-taught fossil hunter found part of a skull on the beach in Lilstock in Somerset. Geologists estimate it came from an ichthyosaur that was 26m (85ft) long – longer even than a blue whale. Most discoveries are much more modest, and much easier to find in any of the many good sites around the British Isles.

Fossils are the naturally preserved remains of an organism that lived in the geologic past; they can be either **trace fossils** (footprints, tracks, trails and burrows), or **body fossils** (a complete or part of a plant or an animal). The British coastline is a great place to look, as erosion and cliff falls constantly uncover new specimens, and coastal currents bring specimens up onto beaches. You can increase your chances of making a successful find by following a few simple guidelines. Fossils do not exist in volcanic rocks, and they very rarely survive the heat and pressure of metamorphic rocks, so your search should begin in sedimentary strata.

The Cambrian period (542 to 488 million years ago) was the time when marine creatures developed hard shells, therefore large numbers of fossils appear here in the geological record for the first time. In Britain, 500-million-year-old trilobite fossils can be found in Marloes Sands in Pembrokeshire, Wales, along with corals, brachiopods (a clam-like fossil) and more.

The 'fossil forest' near Lulworth Cove contains 140-million-year-old fossilised trees.

Trilobite fossils are typically 500 to 400 million years old, and good specimens can be found in coastal outcrops in Wales and the south-west England.

Where to Hunt for Fossils

No matter where you are, you are never that far from a good place to hunt for fossils, so here are some suggestions:

- **Scotland**: The boulder beds on the foreshore at Helmsdale in Sutherland are rich in plant, reptile and fish remains (and the occasional ammonites) from the Jurassic period. You do not need to walk far to find them but take care not to disturb birds nesting in summer.

- **Northern Ireland**: Reptile remains have been discovered along the Antrim coast, but you are more likely to find ammonites and other shells, as well as belemnites (squid-like creatures). Fossils are found on the foreshore, especially after heavy storms. The stones on the beach can be slippery, so this is not a site suitable for young children.

- **Wales**: Abereiddy is great for graptolites. Llantwit Major is the best place for Jurassic fossils, with a wide variety of corals, giant brachiopods (shellfish) and gastropods (sea snails). Penarth is rich in fossils, making it the most popular location in Wales, although the location is becoming over-collected, but you should still come home with some finds.

- **Northern England**: Pterosaur and dinosaur footprints have been found in Robin Hood's Bay in North Yorkshire, and ammonites and reptile fossils are found on the foreshore at nearby Port Mulgrave, especially after storms. The walk down to the beach is not suitable for young children.

- **Eastern England**: The rocks of Hunstanton in Norfolk are rich in fossils of fish, sharks, echinoids (sea urchins) and shells. The foreshore at low tide at Pakefield in Suffolk reveals ammonites, shells and reptile fossils. The boulder clay here collapses after heavy rain, exposing mammal and bird remains. This is a Site of Special Scientific Interest, so check their code of conduct before visiting. Walton-on-the-Naze in Essex is famous for bird remains in the clay, but you can also find shark's teeth and plants. You must not damage the cliff as this is a Site of Special Scientific Interest.

- **Southern England**: The Isle of Sheppey in Kent offers fossilised turtle remains, lobsters, crabs, shark's teeth, snake remains and plants. Take care as you can sink into clay and risk being cut off by the tide. You can also find fossils in the sand in Bracklesham Bay in West Sussex, so there is no need for hammering; on a good day you can find shark's teeth, ray teeth and turtle shells. Spring is the best time when storms have exposed more fossils, but again take care when walking on the soft clay.

- **Isle of Wight**: This is often referred to as 'dinosaur island' with good reason, as it is the richest place in Britain for dinosaur remains. Check out Compton Bay and other bays along the south coast of the island for dinosaur remains and footprints; Bembridge foreland is good for gastropods, brachiopods and other fossils. Yaverland is famous for dinosaur remains (best in late winter at low spring tide), and the beach at Hamstead on the north-west coast offers mammal, crocodile, turtle, crustaceans and fish remains, although access is not easy here and you risk getting stuck in heavy clay and cut off by the tide.

A lucky ammonite find from the Jurrassic coast, Dorset. They are typically 180 to 120 million years old.

Fossil Hunting

Fossil hunting comes with a few simple rules:

- Be responsible and collect only loose fossils – never hammer or dig into the cliff face or ledges.

- In some sites collecting is not allowed at all, so always check first.

- Do not loiter below unstable cliffs.

- Stay safe and follow the beach safety guidelines at the end of chapter 2.

PART 2

THE LIVING COAST

Fish often rise to the surface in the evening, giving herring gulls an opportunity for supper.

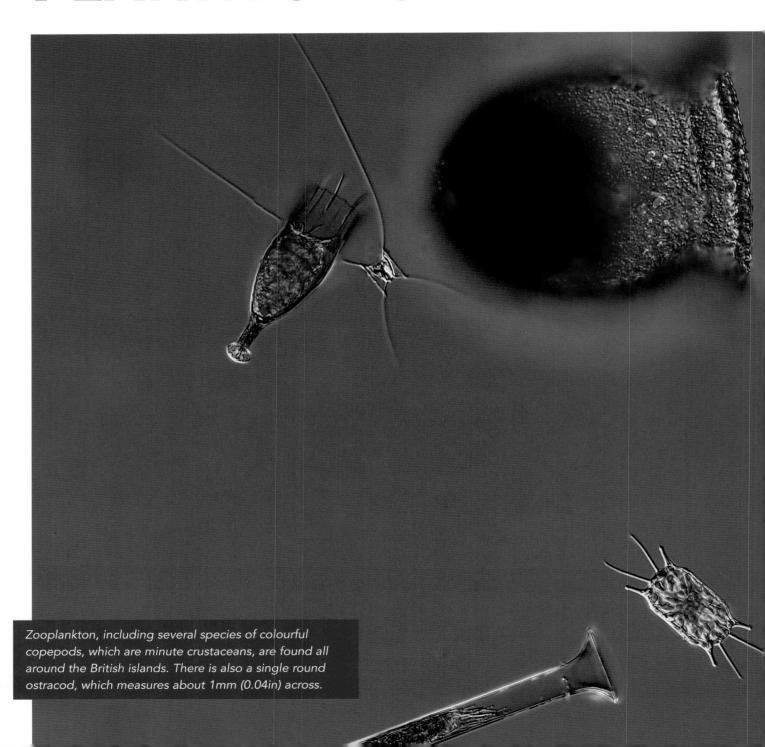

CHAPTER 6

PLANKTON & PLANTS

Zooplankton, including several species of colourful copepods, which are minute crustaceans, are found all around the British islands. There is also a single round ostracod, which measures about 1mm (0.04in) across.

A 1965 a survey in England, Wales and Northern Ireland identified 5,378km (3,342 miles) of British coastline as 'pristine land for permanent preservation'. Today, 94 percent of this land is protected for the future, providing a secure environment for coastal plants and animals. Scotland has its own system, with more than 200 coastal nature conservation areas. It is not a time to be complacent and more can always be done, but it is a good beginning to safeguard the coastal habitat for future generations.

These protected areas are essential if the waters around our islands are going to thrive. The marine ecosystem is complicated – tamper with one area, and the effects elsewhere can be unpredictable. One example is overfishing. Catching fish is not inherently bad for our coastal waters, but if a particular species is taken faster than stocks can replenish, the consequences can be serious: it affects the way fish reproduce and the speed at which they mature, and it can lead to an imbalance up and down the marine food chain.

THE COASTAL FOOD CHAIN

Most marine life in the ocean is found around the coast, where rivers sweep nutrients into the sea to provide a constant supply of food in a complex system called the **food chain**. Plant and animal life in the sea can be divided into three main groups: the largest by far are the **producers**, found at the bottom of the food chain. The producers are mostly plants that use carbon dioxide, nutrients and the Sun to create the energy they need; they are the equivalent of wheat, rice, fruit and vegetables on land. Further up the food chain come the **consumers**, which derive their energy by feeding on producers and other consumers. They include everything from tiny zooplankton that graze on plants and phytoplankton, to large predators like sharks, whales and dolphins. Finally, there are the **decomposers** (whose importance is often overlooked), which break down dead and decaying matter and recycle the energy back into the food chain. This fascinating ecosystem is working overtime all around our shoreline.

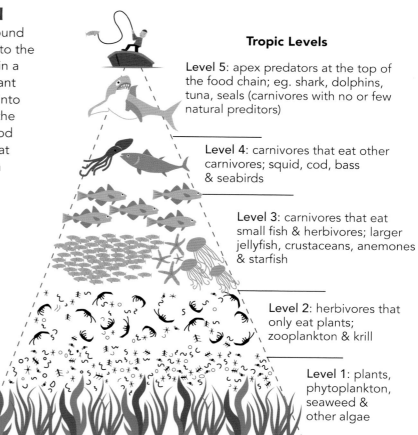

Tropic Levels

Level 5: apex predators at the top of the food chain; eg. shark, dolphins, tuna, seals (carnivores with no or few natural preditors)

Level 4: carnivores that eat other carnivores; squid, cod, bass & seabirds

Level 3: carnivores that eat small fish & herbivores; larger jellyfish, crustaceans, anemones & starfish

Level 2: herbivores that only eat plants; zooplankton & krill

Level 1: plants, phytoplankton, seaweed & other algae

The marine food chain or pyramid.

At each step up the five levels in the food pyramid, the animals become larger but there are fewer of them. In terms of total biomass, there are 10 times more phytoplankton (producers) than there are zooplankton, and this ratio continues all the way up the five trophic levels. Energy is lost in the food system by predators through movement, heat generation, excreting waste, and so on. This means that top-level consumers such as sharks, dolphins and seals are supported by billions of smaller animals and trillions of primary producers lower down the food chain.

Dead creatures and animal parts not eaten by predators sink to the bottom, where they are devoured by the decomposers. Decomposers include bottom-dwelling scavengers like starfish, crabs, lobsters and some fish; organic material is also decomposed by bacteria, with the resulting waste being fed back into the system as nutrients. When a large creature such as a whale dies, an entire ecosystem pops up to consume the new source of food. Because decomposers feed on decaying organic matter, they often experience higher rates of contamination and toxins from pollutants. Crabs, and especially their larvae, seem particularly vulnerable to pesticides.

PHYTOPLANKTON

The hierarchy in the food chain starts with phytoplankton (microscopic plants) which inhabit the sunlit layer of the ocean. (Incidentally, the word plankton is derived from the Greek planktos, meaning 'wanderer' or 'drifter'.) Phytoplankton are essential to the wellbeing of our coastal waters – if you cut off the sunshine or nutrients, this primary source of food simply dies out.

These simple **algae** are primitive plants capable of **photosynthesis**, where light energy is converted into chemical energy, and carbon dioxide and water are converted into organic matter. Because all algal growth depends on sunlight, they are generally found within 20m (66ft) of the sea surface. Phytoplankton have limited movement, and generally drift with the currents, close to the surface. It is estimated that 80 percent of all the oxygen on Earth is produced by these tiny marine organisms. The oldest algae-like marine plants

are 1 billion years old; by comparison, the first land plants did not appear until about 470 million years ago.

Around the British coastline, most phytoplankton are either tiny **diatoms** or the slightly larger **dinoflagellates**. If you are lucky, you can sometimes see **bioluminescence** in the water which looks like sparkling lights; this is usually caused by dinoflagellates, and it is thought the flashes are a response to confuse and deter predators.

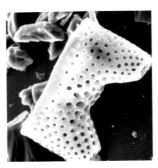

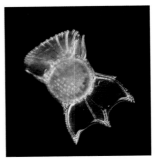

A diatom (left) just 0.05mm (0.0012in) across.

A dinoflagellate, about 0.01mm (0.0024in) from top to bottom.

Dinoflagellate bioluminescence in a coastal bay.

Dinoflagellates and other phytoplankton will sometimes reproduce in abnormally large numbers and create red tides, although the water can also be brown or green. These conditions occur naturally but are more likely to arise from excess nutrients entering the sea from agricultural fertilisers or sewage treatment. Red tides produce harmful toxins which can kill fish, so be wary of eating local shellfish as filter feeders concentrate the toxins, and swimming can also have painful consequences. A red tide occurred off Charleston and Porthpean in Cornwall in October 2019, and several people who were in the sea at the time claimed they developed a rash and sore patches on their skin.

A dinoflagellate red tide on the coast.

Algal growth can also create **sea foam**, which is usually seen in the surf zone or blowing up the beach. It looks as if somebody has poured a box of washing detergent in the sea, but this is a perfectly natural phenomenon. When large blooms of algae decay offshore, sizeable quantities of decomposing organic material can wash ashore, creating the foam. Even though it looks unsightly, most sea foam is not harmful, and usually indicates a productive ocean ecosystem.

Although sea foam looks unsightly, it is a perfectly natural phenomenon and usually does not have any serious health hazards.

SEAWEEDS

There are around 700 species of seaweed around our coastline, making them the most obvious and prolific marine algae. Seaweeds come in a variety of bright colours which helps identify them:
- **Brown** is usually kelp and wrack
- **Red** tends to be carrageen, dulse and moss
- **Green** is usually sea lettuce and spongeweed

Seaweeds reproduce by spores instead of flowers, and their growing season is similar to land plants. They make new growth in spring when the water warms up and days grow longer, and continue to grow quickly throughout the summer, only to wither away in winter.

Kelp is a deep-water seaweed that plays an important role offshore by capturing three-quarters of all carbon absorbed by the ocean. If the water is particularly clear, kelp grows from about 2m (6.6ft) down to depths of up to 45m (147ft) around the British Isles. It grows on rocky surfaces, where their root-like **holdfasts** grip stones, creating great underwater forests. Their leaf-like fronds contain air-filled bladders, which keep the kelp stalks floating upright. British waters have more kelp species than any other European country, with seven out of the 14 European species found in our waters.

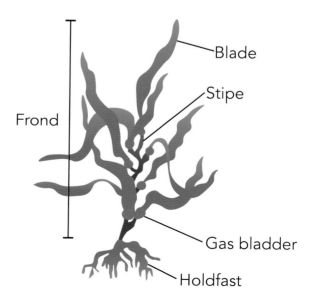

The anatomy of giant kelp (Macrocystis pyrifera) is similar to most seaweeds.

The brown seaweed, bladderwrack, (Fucus vesiculosus) is common all around our coastline; these were found on a beach on the Isle of Arran.

Serrated wrack (Fucus serratus) is another common brown seaweed; it is found in rock pools and often dominates the rocky lower shoreline.

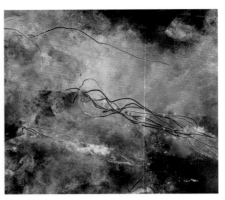

The hairy strands of the brown seaweed, dead man's rope (Chorda filum), grow to 1m (3.3ft); behind is a soft blanket weed (Cladophora glomerata).

Red algae (Rhodophyta), the distinct bright grass-green seaweed gutweed (Ulva intestinalis), and dead man's rope (Chorda filum).

Carrageen (Chondrus crispus), also called sea moss or Irish moss. This abundant red seaweed grows up to 20cm (8in) a year round in tide pools and inlets.

The red seaweed dulse (Palmaria palmata) is a health snack, often found for sale in health food stores or fishmarkets.

Laver (Porphyra umbilicalis) draped over an exposed rock at Osmington Mills, Dorset.

Sea lettuce (Ulva lactuca) grows up to 30cm (12in) throughout the British Isles.

You can find rock pools all around our coastline, and these are ideal places to identify seaweeds.

Edible Seaweeds

Seaweed is found throughout the world's oceans, and none is known to be poisonous. Many species are delicious to eat, and kelp in particular has become a new health food craze. All seaweeds are rich in iodine and calcium, and packed full of minerals, amino acids and micro-nutrients.

The best time to harvest seaweed is in May and June. They attach to rocks and stones with a holdfast, and from here it grows a stem, called the **stipe**. It is important not to damage the algae when harvesting and leave plenty of space around the stipe to allow the seaweed to regenerate. Alternatively, you can gather the free-floating leaves that have been washed free from their attachments.

Collect seaweed from clean coastal areas, away from domestic and industrial outfalls. Wash thoroughly in freshwater to remove sand and debris before cooking. Seaweed has a distinctive flavour, and the taste might take a little time to get used to, although you will almost certainly have eaten some in one form or another, because they are used to thicken soups, emulsify ice-cream and make sausage skins.

Carrageen or Irish moss grows in clusters on rocks and stones and is best eaten fresh. It is abundant in the Channel Islands, where it was used during the Second World War to thicken soups and stews. Carrageen was in such demand that fishing boats collecting the seaweed, in coastal waters ran the risk of hitting mines.

Laver is common, especially on western coasts, and considered a delicacy in south Wales. This red alga has thin, translucent-purple fronds that grow on rocks and stones; when dry, it looks as if the rocks have been draped in shiny, black polythene bags. Laver is best boiled, then blended to a puree – often served on toast.

Other seaweeds can be cooked like laver or eaten as a vegetable, including the **common sea lettuce** (*ulva latuca*), frequently found where freshwater runs into the sea. **Green laver** (*Monostroma grevillea*) looks similar, though more delicate and less common.

Samphire

Although not a seaweed, **marsh samphire** is commonly found on salt marshes. It is a small, stubby plant that grows to 30cm (12in), with plump, shiny bright green stems; it often grows in beds covering several acres. The best plants are covered at high water, so if you are collecting samphire, you will need a bucket and wellington boots. Mid-June is the best time, and the plant should always be cut and not picked (which destroys the root). Young plants can be eaten raw in a salad; older plants can be boiled lightly, and they have the delicate taste of asparagus. As with all seaweeds, do not collect more than you plan to eat. In the past, it was used in the manufacture of soap and glass, hence its other common name, **glasswort**.

Samphire (Salicornia europaea) takes its name from the French Saint Pierre, the patron saint of fishermen.

SEAGRASS

Seagrasses are common around our shores and grow throughout the British Isles. They require lots of sunlight, so are found in very shallow water off our beaches in water depths of no more than 4m (13ft), where they form grassy beds or 'meadows'. All four species in Britain have long, narrow green leaves. They are the only flowering plants to live in seawater and they pollinate while submerged.

Seagrass is truly a 'wonder plant', able to capture carbon up to 35 times faster than tropical rainforests. Even though these grasses cover less than 0.2 percent of the seafloor worldwide, they absorb an astonishing 10 percent or more of the ocean's carbon, making them significant in the fight against climate change. Seagrass meadows provide food, shelter and nurseries for young fish; they are also home for both our native species of seahorse.

The wonder plant – seagrass (Zostera marina). This grass-like flowering plant has long, narrow, dark green leaves up to 50cm (20in). The plant grows from a creeping rhizome that helps to bind the bottom sediment together.

OTHER COASTAL PLANTS

Tidal marshes and wetlands provide a unique environment for many plants, which have to adapt to being flooded with saltwater twice a day. These plants were covered in detail in chapter 3. Away from the water's edge and up onto the foreshore is the realm of stabilising plants such as marram, sand couchgrass and lyme grass. These were covered in chapter 2.

LICHENS

If you are scrambling around the beach looking at rocks or hunting for fossils, spare a thought for the lichens. These beautiful and unusual organisms are worth getting down on your hands and knees to have a closer look; even better, use a magnifying glass. Lichens come in many colours, sizes, and forms.

The common orange sea lichen (Caloplaca marina) on the foreshore at Largs, Scotland.

Some are plant-like in shape, although they are not true plants. Nor are they a single entity, but an unusual mix of two separate organisms – fungi and algae. The dominant partner is the fungi, which gives lichens most of their characteristics, from their shape to their fruiting bodies. The algae can be either green or blue-green, and many lichens have both types. Lichens are found on rocks, in soil, on tree bark, in dunes and on seawalls, so you are likely to find them anywhere along the shoreline, and even down into the inter-tidal zone.

ZOOPLANKTON

Next level up the food chain are the primary consumers, which feed on algae, plants, microorganisms and sometimes other microscopic animals.

Zooplankton is the collective term for a wide range of marine animals, ranging from tiny protozoans (single-cell organisms) such as foraminiferans, radiolarians and non-photosynthesising dinoflagellates, as well as juvenile fish, small jellyfish, and krill (small shrimp-like crustaceans). They vary in size from 0.05mm (0.002in) across to several centimetres (juvenile fish). Zooplankton respond to light and can swim, so they move to deeper water during the daytime to avoid predators and migrate up at night to feed either on phytoplankton or each other. This is one reason why sea fishing is often better in early morning or evening, as predators follow the food.

Various species of larger zooplankton, measuring up to 1cm (0.4in) long.

CHAPTER 7

FREE DRIFTERS & BOTTOM DWELLERS

The plumose anemone (Metridium Senile) is a common sea anemone in British coastal waters, and arguably the most beautiful.

In July 2018, a group of dolphin watchers off the coast of Ceredigion in west Wales, spotted what they thought was a pink oil slick. As they got closer, it became clear that the colourful pattern on the surface was actually caused by thousands of jellyfish, packed tightly together. These graceful marine creatures have been on Earth for at least 600 million years, certainly long before the dinosaurs, making them the oldest multi-organ animal group. But in recent years, scientists have been recording frequent jellyfish blooms like this one in the Irish Sea, and there is no simple explanation.

It is possible that global warming might be one cause, as the immature jellyfish polyps grow better in warmer water, but other factors are at work in our coastal waters. Overfishing removes fish, leaving more food for jellyfish. Animal, sewage and farm waste flowing into our coastal waters causes algal blooms, and this too is good news for hungry jellyfish. Whatever the explanation for these periodic explosions in the jellyfish population, this is a good example of how marine creatures – both big and small – constantly adapt to changing conditions.

JELLYFISH

Despite their passive appearance, jellyfish are carnivores and very effective predators. They use their sting to subdue small zooplankton, crustaceans and fish; they also eat eggs and invertebrates that stick to their tentacles. This puts them halfway up the marine food chain, along with a host of other marine carnivores including sea anemones, starfish, small fish and worms.

Jellyfish, sometimes called sea jellies, have a soft, bell-shaped body and long, stinging, tentacles. A jellyfish is 97 percent water, and they move by contracting their body in pulses; they have little control over direction, and generally drift with the current.

These dome-shaped invertebrates are found all over the world, from surface waters to the deep sea, and there are six species of true jellyfish in British waters. The biggest is the barrel jellyfish, and a magnificent specimen 1.5m (5ft) across was spotted off the coast of Cornwall in July 2019.

Jellyfish Stings

Most species found in British waters only deliver a mild sting; a reaction can develop immediately or be delayed for several hours.

Small children are particularly at risk if they put a jellyfish or a portion of a tentacle in their mouths; any sting to the mouth or throat can cause sudden and severe swelling, which is potentially very serious. A severe allergic reaction (anaphylaxis) may require emergency care.

The best treatment for minor stings is to get out of the water as quickly as possible and rinse the affected area with vinegar or a commercial spray if available.

Carefully remove any tentacles still attached to the skin; a heat pack or soaking in warm water for 40 minutes also helps.

The waters around the British Isles are gradually becoming warmer due to climate change, and jellyfish are on the increase. If you see them collect in large clusters known as blooms, it is advisable to leave the water as soon as possible.

Jellyfish prefer warm water and are most likely to be seen around the British Isles from mid-spring through to early autumn, depending on the species and the weather and water conditions. When conditions are favourable, jellyfish swarm in vast numbers off our beaches; hundreds of people are stung every year.

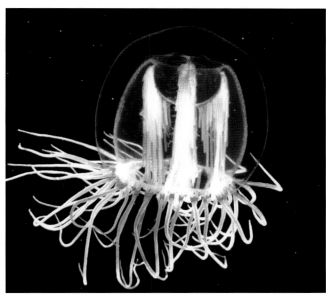

A tiny planktonic jellyfish seen off the coast of the Channel Islands; they move by contracting their muscles, creating pulse-like movements. The tentacles contain harpoon-like stinging cells to capture small planktonic shrimp and crabs.

The barrel jellyfish (Rhizostoma pulmo) is the largest species likely to be found around our coastline and can grow to 90cm (3ft) or more, and weigh in excess of 35kg (77lbs). They have eight frilly arms with small stinging tentacles.

The common or moon jellyfish (Aurelia aurita) lives on plankton, and their sting is relatively harmless. It is almost translucent and grows to 40cm (16in) across. The jellyfish can be easily identified by the horseshoe-shaped gonads on its bell; these are used in sexual reproduction.

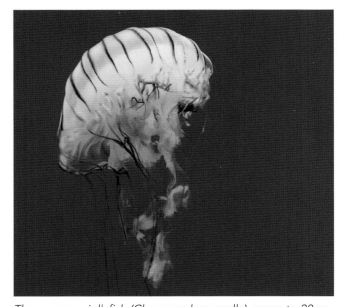

The compass jellyfish (Chrysaora hysoscella) grows to 20cm (8in) across or more. It is common in summer, and feeds on small fish, crustaceans and other jellyfish. Their distinct brown radial pattern on their bell resembles a compass. They can give a painful sting.

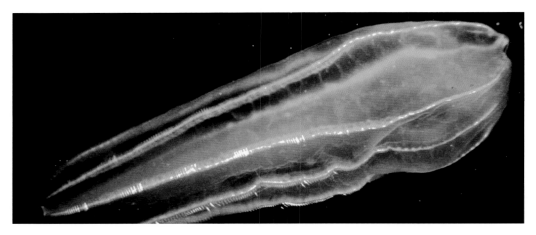

The carnivorous comb jelly (Beroidae) is not a true jellyfish, but a member of the ctenophore family. They use hair-like projections called cilia to move through the water and they range in size from several millimetres to 1.5m (5ft). Most comb jellies give off bioluminisence, especially when disturbed.

Perhaps the best known and most feared 'jellyfish' in British waters is the Portuguese man o'war, which is not a true jellyfish, but a **siphonophore**. These are not individuals, but a group of carnivorous animals called **zooids** which grow within their colony, and cannot live independently; instead, they work together and function as a single entity. All zooids within a colony are derived from the same embryo and are therefore genetically identical. The colony has a gelatinous texture, and it usually disintegrates when caught in a net, making it difficult for marine biologists to study.

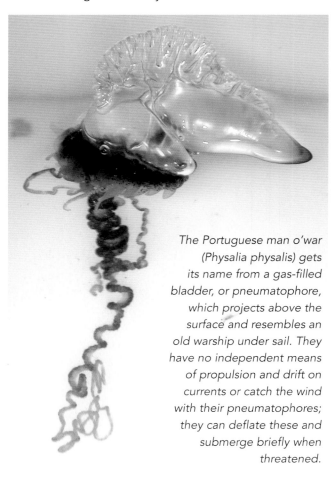

The Portuguese man o'war (Physalia physalis) gets its name from a gas-filled bladder, or pneumatophore, which projects above the surface and resembles an old warship under sail. They have no independent means of propulsion and drift on currents or catch the wind with their pneumatophores; they can deflate these and submerge briefly when threatened.

SEA ANEMONES

If you want to see something truly exotic and flamboyant around the British coastline, put on a mask and snorkel and take a close look at sea anemones. If you prefer to stay dry, marine aquariums can also give you a great view of these amazing creatures.

Given their striking colour, it is no surprise that they are related to corals and jellyfish and are named after the equally flashy terrestrial flower. There are more than 70 species around our coastal waters. They have adapted to a wide range of habitats, from the muddy depths of sea lochs to the seashore, where they are commonly found in rock pools; they also live on wrecks and offshore reefs.

The upper part of the anemone's cylindrical body is hollow, and its mouth is surrounded by seductive, waving tentacles – but beware, for they contain a toxin that will sting any unfortunate small fish that comes within range. (The sting can be unpleasant if you touch them.) Once its prey is subdued, the anemone uses its tentacles to grasp the food and sweep it into its mouth. These animals also filter feed on plankton and larger anemones take small crabs, starfish and jellyfish.

A beadlet anemone (Actinia equina).

Sea anemones may stay in the same place for days or even weeks. If they do move, it is agonisingly slow, and really needs to be captured by time-lapse photography. They can also inflate and release themselves, letting the current take them to a new location, or gently swim by moving their tentacles; sometimes they even hitch a ride by attaching themselves to other sea creatures.

Most anemones reproduce asexually through budding, where fragments break off and develop into new individuals; others stretch themselves along their base and split across the middle, resulting in two new anemones of equal size.

Anemones can also develop a symbiotic relationship with hermit crabs: the anemone moves with the hermit crab, allowing it to catch more food; the hermit crab gets protection because the sea anemone's stinging tentacles scare away predators.

The appropriately named snakelocks anemone (Anemonia viridis). They inhabit the southern and western shores of the British Isles and are usually found in sunny rock pools.

SEA SQUIRTS

Sea squirts are another filter-feeding invertebrate found around our coastline. They start life as tadpole-like larvae; once they find a suitable place to settle down, they digest their own brain, tail and spine-like structure, before metamorphosing into adults. Sea squirts are barrel-shaped filter feeders, and take in seawater using one siphon, from which they filter plankton and detritus; the waste water is then expelled from a second siphon (hence their common name). There are many different types of sea squirt found in UK seas; some are solitary individuals, but many live together in colonies.

The light-bulb sea squirt (Clavelina lepadiformis) grows in colonies and is common around much of the UK. They grow up to 2cm (0.8in) long at depths of up to 50m (164ft). They are also often found in harbours and marinas.

ECHINODERMS

The scientific name Echinus comes from the Greek word for hedgehog, and this group includes bottom-dwelling, hard-shelled **sea urchins**, **starfish**, **brittlestars** and **sea cucumbers**.

Sea urchins have a hard, round shell, and move along the seafloor by crawling on their tubular feet or pushing themselves with spines that protrude from their shell. They live on hard and rocky seabeds throughout the British Isles. They are active predators and feed through their mouths, which are positioned on their undersides. They graze using a beak-like mouth to scrape algae and small invertebrates off hard surfaces. Their predators include sea otters, starfish and humans.

The largest of the five species found in British waters is the **edible** or **common sea urchin**, which grows to 15cm (6in). It is often pinkish purple, but can also be red, green, yellow or white. It lives on the seabed down to depths of 40m (131ft) and is occasionally found in rock pools on a low tide. This sea urchin feasts on seaweeds, small worm-like bryozoans, barnacles and anything else it can find.

A common or edible sea urchin (Echinus esculentus) in a Scottish loch.

The **green sea urchin** is a smaller cousin, about 5cm (2in) across. It lives amongst seaweed on rocky shores to a depth of about 30m (98ft), grazing on seaweed and sponges; it also eats barnacles, mussels and worms. Their strong short spines are often tipped with a brilliant purple, giving them their other name of purple-tipped urchin.

The third species found in British waters is the **purple sea urchin**, which spends its life buried in sand and gravel, and is only rarely exposed on very low tides. It grows to around 6cm (2.4in) and is considered by some to be the tastiest of all sea urchins. Only a small part is eaten – the gonads or sex organs, which run along the inside of the shell. Demand in the UK has increased in recent years and their exploitation is unregulated; this has led to a significant fall in numbers. They are now threatened in British waters, and they could soon become an endangered species.

The green sea urchin (Psammechinus miliaris) is found all around the British coastline, but is scarce in northern North Sea.

The purple sea urchin (Paracentrotus lividus) is found in the Channel Islands, along the west coasts of Ireland and Scotland, and in south-west England.

Sand dollars are the skeletons of flattish urchins and can often be found washed up on the beach. The markings show the characteristic pentamerism (meaning five similar parts) of sea urchins and starfish.

The beautiful **starfish** is not a true fish but is closely related to the sea urchin. More than 30 species are found in British waters, on rocky shores and in rock pools: many are predatory, and some even eat other starfish. The **common starfish** (*Asterias rubens*), for example, feeds on mussels and other molluscs by prizing open their shell with its strong arms. Once the shellfish is weakened the starfish opens the shell and inserts its stomach into the shellfish, where it digests the soft flesh inside – it effectively feeds with its stomach outside its body.

Starfish can be quite active and move using their hollow tube arms, each with a small sucker at the end. If a starfish loses an arm, it will grow another; sometimes a starfish will also shed an arm which is damaged, or to escape predators.

Most starfish breed by releasing eggs and sperm into the open water, where fertilisation occurs; the larvae then drift freely as zooplankton, allowing them wide dispersal. Other species lay eggs on the shore, which hatch into tiny starfish.

A common starfish found high and dry on a sandy beach, making it very vulnerable to hungry gulls.

The distinctive spiny starfish (Marthasterias glacialis) grows to 65cm (25.6in) and is commonly found along the west coast from Devon to the north of Scotland.

Brittlestars are closely related to starfish and have five very thin arms, which break easily if handled. Each arm joint has small spines, giving the creature a decidedly spiky look. They move quickly by wriggling their flexible arms in a snake-like movement – hence their other common name, serpent stars. Fossils of brittlestars have been found in rocks more than 500 million years old.

Many other groups of echinoderms can be found around our shores, which go under the popular names of **sea cucumbers**, **sea gherkins**, and **cotton spinners**. Sea cucumbers are so called because their elongated body resembles the fruit; they have thick skin with no visible sensory organs, and only a limited ability to move. The **gravel sea cucumber** is the most common species in British waters and found mostly along the west coast of Britain and Ireland. It lives on coarse sand, gravel or shingle seabeds down to about 100m (328ft).

The common brittlestar (Ophiothrix fragilis) has a body up to 2cm (0.8in) in diameter, and its five spiny arms are about five times its disk diameter. They are usually brown or grey, but range through purple, red, orange, yellow and white. Found in crevices and under boulders in rock pools.

The gravel sea cucumber (Neopentadactyla mixta) has an elongated body up to 15cm (6in) long, with small feet underneath. The creature buries into the bottom sediment, then leaves its tentacles exposed to catch organic matter as it drifts past.

The cotton spinner (Holothuria forskali) is an unmistakably large sea cucumber, growing to 25cm (10in), and found in rocky crevices and on lower rocky foreshores. When disturbed it ejects a mass of sticky cotton-like threads which entangle the would-be predator.

MARINE WORMS

They might not be the most glamorous of creatures, but the importance of worms to the marine environment should never be underestimated. Some species play a key role in cleaning up and consolidating mud- and sandflats, whilst others are opportunistic filter feeders; they all provide an important source of food for fish, molluscs and birds.

The species found in British waters can be divided into three simple groups: **flatworms**, **ribbon worms** and **segmented worms**.

Flatworms are only a few centimetres long, and easily overlooked. They are leaf-shaped, with a flattened body and simple digestive system. There is a single opening under the body which serves as both a mouth and an anus. They glide slowly around the rocks using cilia (microscopic hairs) under their body as miniature paddles. Flatworms are carnivorous and feed on small, sedentary organisms. With care, you can find them under stones, in kelp holdfasts and feeding on sea squirts (one of their favourite foods).

Although still quite primitive animals, the long and thin **ribbon worms** are slightly more advanced than flatworms. For example, they have a more efficient digestive system with both a mouth and an anus. They also have a circulatory system and a proboscis (a trunk-like extension near the mouth), which they use to trap prey and use in defence. The **bootlace worm** is the longest animal on Earth, and grows to more than 30m (98ft). When the worm is irritated, it releases large amounts of thick mucus which contains a neurotoxin, capable of killing a crab.

As their common name suggests, **segmented worms** have long, segmented bodies, sometimes covered with soft plates or hairs; other have bristles on the segments. Their gut runs the full length of their body from the mouth at one end to the anus at the other. These are the most advanced of the marine worms, and have developed internal organs and a nervous system. Many segmented worms move around freely, but usually burrow into the soft seabed or remain hidden amongst organic debris, so they are rarely seen. Those which are most visible are the species that live in a tube, which they construct for protection. In these species, their head often protrudes from the opening of the tube, frequently with feathery tentacles used to catch food and for respiration. Some species, such as the **fan worm** or the **peacock worm**, have very striking tentacles.

The candy stripe flatworm (Prostheceraeus vittatus) is the best looking of all the species, and grows to 5cm (2in). They are found mainly in the south and west of the British Isles.

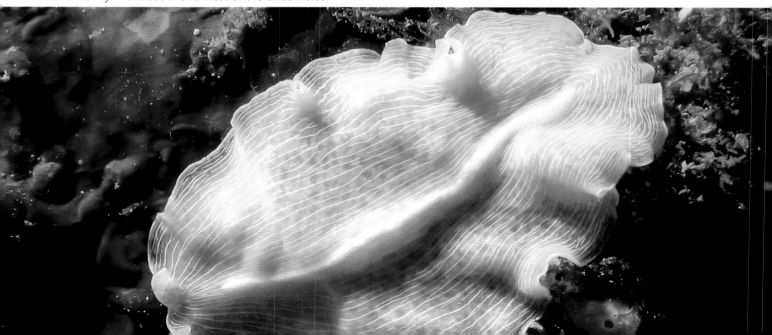

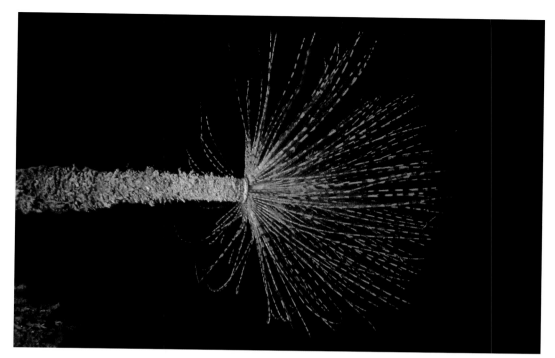

The European fan worm (Sabella spallanzanii) is found around the British Isles in both shallow and deeper water, growing to 40cm (16in). They have stiff, sandy tubes which are formed from hardened mucus secreted by the worm. A large female can produce upwards of 50,000 eggs during the breeding season.

The fan worm (Bispira volutacornis) has more than 200 tentacles, and the fan usually retracts into the tube when disturbed. The worm is widespread in the south-west of England, with occasional sighting off Flamborough Head, the Isle of Man and western Scotland.

Another common segmented worm is the **bristle** or **polychaete worm**. These include the **lugworm**, which lives in silty sand, and feeds on tiny animals and dead matter that it filters through the sand. The body of the lugworm has a rounded thick main section, with bristly gills running down one side, and a much thinner tail section. Lugworms are usually brownish, although in some regions they can be reddish, or much darker to the point of being almost black, with the tail section being a lighter yellow colour. Generally, smaller worms live nearer the high-tide mark, with individuals become progressively bigger towards the low-water mark.

The lugworm (Arenicola marina) rarely exceeds 13cm (5.1in) and is a popular fishing bait.

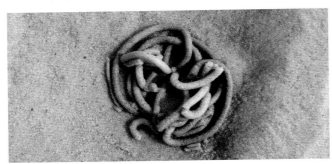

The U-shaped burrows of the lugworm are easily identified by tubular 'casts' on the foreshore; this is material that they excrete as they digest their food.

Marine worms are an important source of food for other coastal creatures. Here an oystercatcher (Haematopus ostralegus) enjoys a meal of fresh lugworm.

The **ragworm** is another common segmented worm, and several species are commonly found around the British coastline. Its body is composed of scores of segments, and is sometimes flattened, depending on the species. These are active predatory worms which burrow in mud and silt where they consume tiny creatures; they also feed on microscopic organisms suspended in seawater, and on organic plant matter. They are particularly prized as fishing bait.

The king ragworm (Alitta virens) usually grows to 30cm (12in) long, although the largest specimens can reach 90cm (3ft). The adult has up to 200 flattened segments, a green metallic sheen and four tentacles each side of its head.

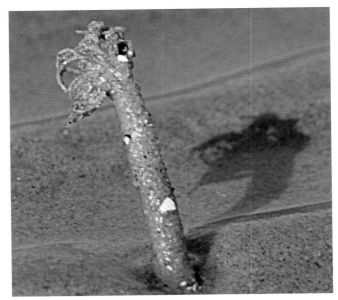

The sand mason worm (Lanice conchilega) is a species of burrowing polychaete worm. It builds a characteristic tube of cemented sand grains and shell fragments, which projects from the seabed. The tube often has ornate branches made from sand and shell which are used to catch food. The tops of the tubes can be spotted on beaches at low tide. They can be found as single individuals, or in dense beds of thousands.

CHAPTER 8

MOLLUSCS & CRUSTACEANS

The common octopus (Octopus vulgaris) generally lives in shallow, rocky areas and is found mainly in the south and west of the British Isles.

In 2008, an octopus called Paul from the Weymouth Sea Life Centre in Dorset correctly predicted four of Germany's six results in the Euro 2008 soccer tournament. Having become an international celebrity, the octopus moved to an aquarium in Germany where he again forecast the results of seven football matches involving Germany at the 2010 World Cup. These molluscs rarely live more than two years, and sadly Paul – the 'psychic' octopus – died from natural causes later that year, aged two and a half. His memory lives on in the form of a 2m (6.6ft) plastic likeness clutching a giant soccer ball; inside is a small golden urn containing his ashes.

Whatever skills Paul seemed to possess, they did not include telepathic clairvoyance. But to his credit, octopuses are well known for being highly intelligent, and their brain-to-body ratio is the largest of any invertebrate, and larger even than many vertebrates (although not mammals). In controlled experiments, they have solved mazes and completed tricky tasks to get food rewards. Most aquariums that keep octopuses tell tales of their exploits, including overnight raids into neighbouring tanks for food, and even learning to turn off lights by directing jets of water at them, causing an electrical short-circuit.

MOLLUSCS

It is difficult to know where to start when describing marine **molluscs**, because this group is so large and diverse that they seem to have little in common with each other. Worldwide, they account for 23 percent of all marine creatures. All molluscs are invertebrates, which means they have no backbone, and are closely related to terrestrial snails and slugs.

Some molluscs are herbivores, others are carnivores; some are filter feeders whilst others are predators; some have one shell, others have two; some carry their shell inside their bodies, whilst others have no shell at all. The smallest are less than 1mm (0.04in) across, and the largest can grow to 13m (43ft).

This extraordinary group of animals range from the limpet that grazes on algae, to the octopus which has excellent eyesight and hunts aggressively for its prey.

GASTROPODS

Sea slugs are gastropods (i.e. snails) that have lost their shells during evolution, or have a greatly reduced or internal shell. Most of these molluscs are partially translucent, but have also developed bright colours as a warning of their unpleasant taste, as they also protect themselves by emitting an obnoxious secretion. Like all gastropods, they have small razor-sharp teeth, called **radulas**.

The **common limpet** used to be easy to find when scrambling around the shoreline, but are now mostly found in northern England and Scotland, having been nudged out by invading species. The limpet has a cone-shaped shell and a fleshy light orange body, which is edible. Its lower body consists of a large 'foot', used to clamp onto rocks. It feeds by slowly crawling over rocks and grazing on algae. Limpets are found in the inter-tidal zone or in shallow water to about 10m (33ft). There are around half a dozen less common species around our shores which have similar shaped shells.

The sea slug (Elysia viridis) grows to 3cm (1.2in), with a smooth, bright green or brown body and iridescent spots; it has two wing-like flaps which extend along its sides and can fold back. They are found in the inter-tidal zone all around the British coastline to a depth of about 5m (16.4ft).

The common limpet (Patella vulgata) clamps itself to the rock tenaciously when disturbed.

The invasive **slipper limpet** is another species, and technically a sea snail rather than a limpet. It comes from North America, and probably arrived in consignments of oysters and mussels in the early twentieth century. Today they are found throughout much of southern England and south Wales. It is thought that the colder northern waters might prevent the species from spreading, although ocean warming could change this in future. They are classed as an invasive species as they smother oyster beds and out-compete native shellfish species for food (including the common limpet). Since 2015 it has been an offence to release slipper limpets into the wild.

The head, mouth and tentacles of the limpet are clearly seen on the underside.

The invasive slipper limpet (Crepidula fornicata) has an arched, rounded shell which resembles a slipper.

Loosely related to the limpet are **winkles, whelks** and **top shells**. These molluscs have a distinctive head, eyes and tentacles, and a rasping tongue-like radula used to feed on rocky surfaces. The common whelk is the largest sea snail found in British waters and can grow to 10cm (4in); it usually lives below low water. This carnivorous mollusc feeds on worms and other molluscs, often using the edge of its own shell to prize open the shells of its victims.

The common periwinkle (Littorina littorea) generally live in colonies. They are found throughout the British Isles and are edible.

The common whelk (Buccinum undatum). The female lays thousands of eggs under rocks, which are frequently washed up on the beach.

Many single-shelled molluscs have evolved patterning on their spiral shells. The pelican's foot (Aporrhais pespelecani) grows up to 4cm (1.6in).

The pointed auger shell (Turritella communis) grows to 5.5cm (2.2in).

The painted top shell (Calliostoma zizyphinum) is found on or below the low-water mark, along sheltered rocky shores; it grows 3cm (1.2in).

The smaller rough periwinkle (Littorina saxatilis) grows to 1.7cm (0.7in) and is found higher up the foreshore. They are herbivores and feed by scraping algae from rocks.

There is a huge diversity of sea snails around the coast of the British Isles, and their empty shells are always being washed up onto the foreshore. Some species can be viewed alive on the beach above low water, usually hidden amongst seaweed. Others can be found in rock pools; a snorkel and mask will give you a fascinating insight into a hidden world below the surface.

BIVALVES

A major group of molluscs are bivalves, which include **oysters, cockles, scallops, clams, mussels** and many others. These molluscs have two hinged shells (called valves), which can be symmetrical or asymmetrical; the majority are filter feeders. Bivalves have several advantages over gastropods which aid their survival: most bury themselves in sediment where they are relatively safe from predators, and scallops can even swim.

Bivalves are popular seafood and have been part of the diet of coastal communities for centuries. **Oysters** were cultured by the Romans and continued to be a staple food for several centuries after they left. These bivalves were so plentiful in the nineteenth century, that the classic Victorian dish of beef and oyster pie was considered a poor man's meal. How times have changed!

The irregular, asymmetric oyster shell is made from flakes. As the mollusc gets older, more flakes are added, and the shells become thicker. Oysters are found on muddy, gravelly beds, and grow to 9cm (3.5in). The native oyster declined to critically low stocks in the first half of the twentieth century due to overfishing and environmental stress. Sustainable fishing has allowed the native oysters to recover, and they are almost always dredged from the wild.

The bigger Pacific oyster comes from Japan, where it has been cultivated for hundreds of years. It was introduced into British waters in 1965 to replace low stocks of the native oyster. Pacifics are almost always farmed and take 18 to 36 months to mature (half the time of the native); they can grow to 20cm (8in) or more, making them commercially more viable. Today, Pacifics represents over 98 percent of the world's harvested oysters.

The Pacific oyster (Magallana gigas) is a highly favoured seafood, usually eaten raw with a squeeze of lemon. Mature specimens can vary from 8 to 40cm (3.1 to 16in) long.

The native oyster (Ostrea edulis) has a mother-of-pearl coating on the inside of the shell, although they do not produce pearls of any size or commercial value.

Pearls

Even though almost all molluscs with shells can secrete pearls, most have no value. Commercial pearls come from pearl oysters, which are not closely related to true oysters, and none live in British waters. Freshwater mussels also yield pearls of commercial value.

A much cheaper seafront delicacy are **cockles**, which have two identical shells hinged at the narrowest point. They live buried in sand or silt around much of the British coastline, and their shells are frequently found washed up. They have a single 'foot' which they use to 'jump', by bending then suddenly straightening it out. There are half a dozen species along our coastline, including the prickly cockle and the rough cockle.

The rough cockle (Acanthocardia tuberculata) grows to about 9.5cm (3.7in), with crenulated margins and between 18 and 20 strong radial ribs. They are usually pale brown with alternating darker concentric bands. They are found from low tide down to 200m (660ft).

The common cockle (Cerastoderma edule) is the familiar edible cockle. The shell has between 22 and 28 radiating ribs, crossed by conspicuous concentric ridges. They grow up to 5cm (2in), but are usually smaller.

The thick trough shell or surf clam (Spisula solida) grows to 5cm (2in). The brownish or yellowish-white shell shows fin concentric lines. The mollusc is found at scattered locations around the coasts of Britain at depths of 5-50m (16-164ft).

Scallops are one of the most prized shellfish in British waters. The king or great scallop grows to 15cm (6in), and the smaller queen scallop to 9cm (3.5in). The variegated, the humpback, and the tiger scallop are smaller species and are not commercially viable.

The prickly cockle (Acanthocardia echinata) has a yellowish-brown shell is up to 7.5cm (3in) in diameter, with 18 to 22 radiating, spiny ridges. The prickly cockle is found throughout the British Isles, and lives within a few centimetres of the sea bottom at depths of 3m (10ft) or more.

The king scallop is harvested using heavy-toothed dredgers. These huge steel frames weigh up to two tonnes and have large teeth that gouge the seabed to dislodge the shellfish; they are then scooped up in a chainmail bag. This the worst possible method of fishing for shellfish because the dredge catches large numbers of other seabed creatures as unwanted by-catch, and badly damages the seabed. Harvesting scallops by divers is much more sustainable.

The king or great scallop (Pecten maximus) is found around most coasts of the Britain Isles, but rarely along the east coast of Great Britain. They prefer areas of clean firm sand or sandy gravel.

The queen scallop (Aequipecten opercularis) is a more ecologically friendly catch. 'Queenies' are active swimmers and leave the seabed in the summer; harvesting them is therefore less destructive.

The smaller humpback scallop (Chlamys varia) is commonly found in the south and south-west. It is about 5cm (2in) across and has up to 70 distinctive radiating ribs.

The hard-shell **clam** or **ocean quahog** is another edible bivalve and grows to 11cm (4.4in). This large burrowing bivalve is found on sandy seabeds around much of the British Isles, and is the longest-lived animal known to man, with one individual found to be 507 years old.

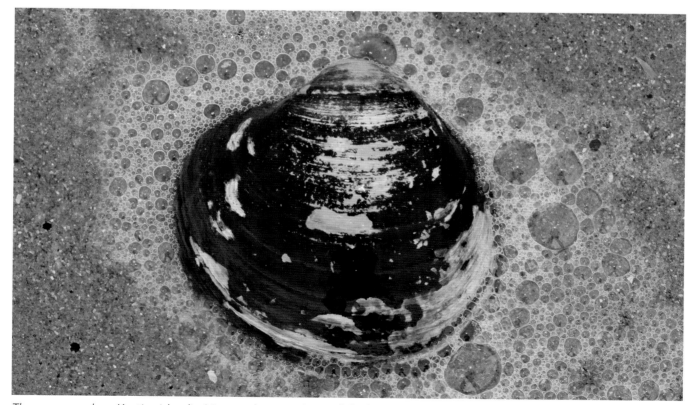

The ocean quahog (Arctica islandica) is very slow growing and the last surviving species of a family of similar clams that date back to the Jurassic period, more than 145 million years ago.

Unlike cockles and scallops, the distinctive blue-black **common mussel** has an asymmetric smooth shell showing sculptured concentric lines, although its shape can vary considerably. These filter feeders are found throughout our islands, where they live along the lower shoreline; they particularly like fast flowing water, where they attach themselves to rocks with fine threads. Mussels are farmed commercially using ropes to gather microscopic mussel spawn. There are large commercial beds in the Wash, Morecambe Bay, Conwy Bay and the estuaries of south-west England, north Wales and western Scotland.

The shell of the bearded mussel (Modiolus barbatus) has distinctive bristles; these small bivalves grow to 6cm (2.4in), and are found mainly along the south and west coasts.

The common or blue mussel (Mytilus edulis) rarely exceeds 10cm (4in), and is extensively farmed.

The **horse mussel** is a larger species, growing to 20cm (8in), and lives below low-water mark. The **bearded mussel** has a distinctive shell with bristles at one end; they too live just below the low-water mark and are only spotted when their shells are washed ashore.

Razor clams or razor shells have long, narrow shells, and are another easily recognisable bivalve. They burrow into sand from the low-tide mark to a depth of about 60m (197ft) using a strong muscular 'foot'. Here they feed on plankton and detritus. Often the only visible part of a living razor clam are two small siphons poking out of the sand. For this reason, you are only likely to see their shells when washed up on a beach, especially after big storms.

The horse mussel (Modiolus modiolus) occurs all around the British Isles but is more common in the north.

Razor clams and other shells washed up on a beach. There are three very similar species along our coastline (Ensis siliqua, E. arcuatus and E. ensis) which are usually found in sandy areas. They grow to 20cm (8in) and can live for 10 years. Overfishing has caused a general decline.

Sometimes you find evidence of mollusc activity rather than the creature itself. The **piddock** is an amazing animal which can burrow into rock (or wood). When they settle on a rock as larvae, they slowly begin to enlarge and deepen their hole as they grow. From here, they filter-feed on organic matter. When the piddock dies, it leaves an empty burrow which often becomes home for other marine species, including small molluscs, juvenile crabs and even small sea anemones. Their long oval shells are distinctively wing shaped, giving piddocks their other common name of angelwings.

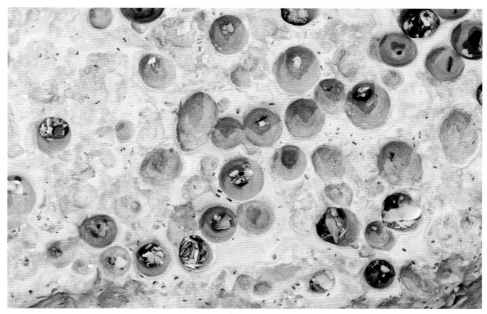

The common piddock (Pholas dactylus) is an unusual clam-like bivalve, which bores through sedimentary rock to create their burrow. They grow to 12cm (5in), and are usually white, but can develop a pinkish colour from filtering red tide algae.

Should you Steal Seashells from the Seashore?

Apart from being a testing tongue-twister, this is also a challenging ethical dilemma. Taking a small bag of shells from a beach is not going to make much of a difference; if a hundred people each take a bag every day, from every beach in the country, for a year… that is when we have a problem.

Under the Coastal Protection Act of 1949, it is against the law to take rocks and sand from a beach – this is intended to make commercial beach extraction illegal. Shells are excluded from the Act, except in Scotland, where it is also illegal to gather shells. However, shells are made from calcium carbonate, an essential building block for marine life, and by collecting them, we are changing the ecosystem. So, what should you do? Well, think first before you do anything, and consider the consequences:

- It goes without say you should not take anything that is still living. Leave it alone to enjoy its short life.

- Ask yourself if you really need the shell. Take only a few, or better still, take lots of photos.

- Spiral shells are used by hermit crabs, and when they need to move house they are vulnerable to predators and the hot sun. If you remove spiral shells, it will take a crab longer to find the right size. These are beautiful shells, but there is somebody who needs them more than you.

- Seagrasses, corals and anemones use shells as anchors – and they are not too choosy which type. Many creatures build their homes on shells, such as barnacles and limpets; other creatures use shells as camouflage or shelter. Some organisms burrow into discarded shells, including some algae and sponges.

- Old shells are used as nest-building materials by several species of birds, and the sand mason worm makes good use of shell fragments to build its tube.

There is no easy answer to this question, except to be sensible and play safe.

Best of all, take nothing but photographs and leave nothing but footprints.

CEPHALOPODS

The animals that break all the rules for being molluscs are the cephalopods, which include **squid**, **cuttlefishes** and **octopuses**. None have an external shell, they are all predatory carnivores, and they all have excellent eyesight and (relatively) large brains – very different from your average bivalve.

Like all cephalopods, **squid** have an elongated body, large eyes, eight arms, two tentacles and a distinct body or **mantle**. They are mostly soft-bodied but have a small internal skeleton, and a beak made from a fibrous material called **chitin**. Cephalopods are fast swimmers, and use jet propulsion to shoot themselves forward quickly, making them formidable hunters. They can also distract predators by ejecting a cloud of black ink, giving them a chance to escape.

Squid use their long tentacles to grab their prey, and their eight arms to hold it firmly, before hacking the luckless creature into small chunks with their beak. Two main species are found in British waters: the **common squid** and the **European squid**. They are more common in the warmer south and south-west, especially in summer and early autumn.

The common squid (Alloteuthis subulata) is the smaller of the two species found in British waters and grows to 20cm (8in). The larger European squid (Loligo vulgaris) is typically 30cm (12in) or more. They gently undulate their side fins when swimming slowly but wrap them around their body when moving quickly using jet-like propulsion.

The closely related common **cuttlefish** also has eight arms and two tentacles, with suckers to hold its prey. They are effective predators with a well-developed brain and good eyesight, but their W-shaped pupils give them a sinister look. They are usually blackish-brown, mottled or striped. They can change colour quickly to merge with their background, either to distract predators or to attract mates. They will sometimes sink into a sandy seabed to hide from predators, leaving only their eyes exposed. They eat small molluscs, crabs, shrimp, fish, octopus, worms, as well as other cuttlefish, and are prey to fish, seals, seabirds and other cuttlefish. Like the squid, they can squirt ink to assist escape.

The common cuttlefish (Sepia officinalis) grows to 45cm (18in) and is the largest of three species found in British waters. They live to depths of 200m (660ft) but move into shallow water to breed in spring.

When cuttlefish die, their chalky internal shells, known as cuttlebones, often wash up on beaches.

The common **octopus** (Octopus vulgaris) generally lives in rocky areas and shallow water around the south and west of the British Isles. They have eight arms, each with two rows of suckers which allow the octopus to taste whatever it is touching. They can also vary and change their colour to adapt to their surroundings.

Octopuses move either by swimming slowly using their arms, or by expelling water from their body, which shoots them forward. This requires a lot of energy and leaves them exhausted which makes them vulnerable to predators. So they generally use jet propulsion only as a last resort. Their powerful arms can prise shells off molluscs, but they also grasp the shell of their prey and poke a toothed protrusion from their mouths to drill into the shell. The octopus secretes a poison which breaks down the muscles and flesh structure of the prey, making it easier for the octopus to force its way into shells. They also feed on fish and scavenge on anything found dead or dying on the seabed. They are prey to larger species such as sharks and other predator fish but use their jet-like propulsion and ink emission to avoid capture.

The common octopus can grow to 100cm (39in). With no backbone, they can squeeze their soft body into the smallest of spaces to avoid predators or hide from their prey.

CRUSTACEANS

Few crustaceans are found on land (woodlice are a rare example), and the vast majority are found in the ocean, including the tiny sand hopper, as well as crabs and lobsters. Not all are scavengers – krill, for example, live in the open ocean and feed on plankton, while barnacles are active filter feeders. However, most bottom-dwelling crustaceans are very effective scavengers.

Globally, there are about 38,000 species, making them the second largest class in the animal kingdom. All have hard skin or a shell, gills and two pairs of antennae. Typically, their bodies comprise of three parts: the head, the thorax (chest) and the abdomen. Most crustaceans hatch from eggs into a larval stage, which look nothing like their adult form. These tiny larvae float near the ocean surface as zooplankton, where they feed and grow. They shed their hard skin several times before becoming adults.

There are hundreds of species of very small crustaceans living around our coastline, so tiny that they mostly go unnoticed. The beach- and sand hoppers, for example, can be found on most sandy beaches in huge numbers. Other species are even smaller, and you will need a magnifying glass to identify them.

The sand hopper (Talitrus saltator), sometimes called a sand flea, grows to 2.5cm (1in). They can bury as deep as 30cm (1ft) and emerge at night to feed on rotting and decomposing marine vegetation.

BARNACLES

These shellfish are confusing. Superficially, a barnacle looks like a small clam or oyster, and they cling to rocks for their entire lives, filter feeding from the high tides. But barnacles are actually crustaceans, not molluscs, and more closely related to crabs, lobsters and shrimp. Barnacles have legs inside their shells, and show the same segmentation as all crustaceans.

Barnacles are so common on our rocky shores that you probably never notice them – until you walk on them in bare feet. There are several different species in the British Isles, and they are difficult to tell apart; the most common is the **acorn barnacle**. They permanently attach to any hard surface, including rocks, pier legs, boats and even other animals. Their body is upside down inside their cone-shaped shell, and when the tide comes in, they open their shell and stick out their feeding legs (called cirri) to catch plankton and other detritus.

The acorn barnacle (Semibalanus balanoides) grows to 2cm (0.8in) and its shell is made from six calcified grey-white plates.

The unmistakable **goose barnacle** grows in dense clusters attached to anything that floats – including ships, ropes and buoys; and they even attached themselves to a chunk of space rocket that washed up in the Isles of Scilly in 2015. Their chalky white shell looks like a small, pale mussel, they grow to 5cm (2in), and feed on plankton and detritus. The barnacle attaches itself to the flotsam with a stem or peduncle, which looks like a long, fleshy neck, up to 50cm (20in) long. They are considered a delicacy in some parts of the world, and in the past ships arriving in England after a long Atlantic crossing had their goose barnacles scraped from the hull and sold as food. A similar species, the **buoy barnacle** (*Dosima fascicularis*), is pale purple and floats on its own spongy 'buoy', rather than being attached to flotsam.

Goose barnacles (Lepas anatifera) attach themselves with a long, dark 'neck', and are most likely found washed up on west and south-west coasts, especially after a storm.

DECAPODS

Further up the scavenger list come the bigger crustaceans – **shrimps, prawns, crabs** and **lobsters**. They all belong to a group of animals called decapods, meaning '10 legs'. **Brown shrimps** are very common around our shores; they bury themselves in sediment in daytime as protection from predators, with only their antennae protruding from the sand; they then emerge at night to feed on plants and any decaying organic matter they find, including dead fish, clams, snails, crabs and worms.

The bigger **common prawns** are found throughout the British Isles, but mainly in the south-west. They have a translucent body with brownish-red tiger stripes and are frequently found in rock pools. They also feed on pretty much anything, with the remains of any dead sea creature at the top of their list. They are omnivores and also feed on plant matter, algae and organic debris.

There are several other species of shrimps and prawns, which live in a similar habitat. All shrimps are commercially valuable species, and they are also an important part of the food chain for a wide range of fish species and marine birds.

The brown shrimp (Crangon crangon) can change colour to camouflage themselves from predators. The overlapping segments can clearly be seen along its back. They usually grow to 6cm (2.4in) but can be as big as 9cm (3.6in).

There is often a lot of confusion between the terms prawn and shrimp. These are colloquial or common terms, not scientific names, and are often used interchangeably; the common prawn found in British waters is actually a shrimp. In practice, the term prawn is generally applied to larger species.

Another confusing common name is the **langoustine**, also known as the Norwegian lobster or Dublin Bay prawn; the meaty tail is called scampi. These are all names for the same crustacean. (The crayfish, which looks very similar, is a freshwater crustacean.) The langoustine (to use its French name) grows to 20cm (8in), occasionally bigger, and is an important commercial catch. They are extensively fished off the west coast of the British Isles, around the Republic of Ireland and in the North Sea.

The langoustine is found at depths between 20 and 800m (66 and 2,625ft), where it burrows into fine sediment which is stable enough to support its tunnel. It emerges at night to feed opportunistically on other crustaceans, molluscs, starfish, and to generally scavenge. Like other crustaceans, the

The common prawn (Palaemon serratus) grows to 11cm (4.3in) and is translucent, with red / brown lines on the carapace and abdomen. They are most common on the west, south and south-west coasts of England and Wales and west coast of Scotland.

langoustine moults once or twice a year, and sheds its exoskeleton to allow it to grow. Mating takes place when the female is still 'soft', and after fertilisation she carries her eggs for eight to nine months. She remains in her burrow all this time, and only emerges to allow her eggs to hatch and the larvae to escape.

A female langoustine (Nephrops norvegicus) with her roe. After mating in early summer, they spawn in September and the female carries her eggs until they hatch the following April or May. The larvae develop as plankton before settling on the seabed six to eight weeks later.

Crouch down and look into any rock pool, and along with brown shrimps, you are most likely to see small **crabs**. If you do not see them at first, look under stones and seaweed, and you almost certainly will. Like all decapods, crabs have 10 legs. The front two develop into powerful claws, used to catch, crush and chop their prey; they are also used for fighting. Crabs have two eyes perched on stalks protruding from their hard shell, and a pair of small feelers between. This makes them ideally suited for scavenging along the seabed. Incidentally, they also have a reputation for being greedy and quarrelsome. Crabs are very common around our shores, with more than 62 species found in British waters.

As crabs grow, a new, soft shell develops before they shed their old hard shell. After moulting, their new shell needs to harden, and this is when crabs are at their greatest risk from predators – and it is also the time when mating occurs. You can tell the difference between female and male crabs by turning them over, where you can see their abdominal flap (which looks like a short tail). Males have narrow flaps, whilst females have wider, rounded flaps, which they use to carry and protect their eggs. The female **common shore crab**, for example, carries about 180,000 eggs

Scampi

When the tails of langoustine are breaded and fried, they are served as scampi. They became fashionable in the 1970s as a pub meal, and they remain popular today. The authenticity of scampi, however, is highly variable. 'Premium' scampi is the whole tail of the langoustine, but these nuggets still only contain about 40 percent of shellfish by weight – the rest is breadcrumbs, bulked out with water. Cheaper scampi has whitefish added – sometimes in significant proportions – and retextured to create the shape of the langoustine tail. The ingredients of 'scampi bites' in one supermarket freezer contained: breadcrumbs 55 percent, minced cod 30 percent, scampi 7 percent, rapeseed oil, plus water. So choose your scampi carefully.

Scampi and chips is a very popular pub meal.

under her abdomen flap. The males fertilise the eggs held by the female and, once they hatch, the larvae become zooplankton, swimming freely in the water. As the larvae develop, they change shape several times before developing into an adult.

There are half a dozen common crabs around the British shoreline, and you can entice them out with a piece of bacon on a weighted crab line. Most British crabs are omnivorous scavengers, and feed on dead animals and plant matter; they will also graze on seaweeds. The larger crabs can crush the shells of molluscs with their claws. They often use one pincer to crush and hold their prey, while using the other to tear small pieces off to eat. Some crabs filter feed by using thin 'hairs' (called setae), which they waft through the water to create a small water current; food particles are then filtered and eaten.

The common shore crab (Carcinus maenas) in a typical aggressive position when threatened. This species is easily identified by the tooth-like projections on the front jagged edge of its shell. They are usually dark green, but colours vary. They increase their body size by 20-33 percent during each moult and it takes about 10 moults to reach a carapace width of 20cm (8in).

You are most likely to find the edible crab (Cancer pagurus) on a fishmonger's slab, as the largest ones live in deep water; they grow to 25cm (10in) across. This reddish-brown crab has a distinctive crimped edge to its shell, looking for all the world like a pink Cornish pasty. © Matthieu Sontag, Licence CC-BY-SA.

A pair of velvet swimming crabs (Necora puber) mating. They look dramatic with their bright red eyes, giving them their other common name, the devil crab. They hide under rocks, and can be aggressive and will snap their claws if threatened. They are typically 5cm (2in) but can grow larger.

A perennial favourite is the hermit crab (Pagurus bernhardus), sometimes called the soldier crab. They are soft-bodied, and do not produce their own shell. Instead, they find an empty spiral shell and move up sizes as they grow. When inside the shell, they use their right pincer to seal the entrance. They are commonly found in rock pools and can grow to 8cm (3.2in). © Thomas Bresson

The Good Crabbing Guide

One of the joys of a family seaside holiday is dangling a line attached to a piece of bacon over the side of the quay to see who can catch the biggest crab. Crabbing can be great fun, and a great way to introduce children to marine ecology. But we also want to make sure that our crabs do not suffer in the process.

- Do not use a line with a hook; tie your bacon to the end of a line, or place in a pouch or a pair of old tights; drop nets are not a good idea, as crabs tend to get caught up in them.

- Use a big bucket with plenty of water and add rocks and seaweed to reduce stress on the crabs.

- Do not put more than six or eight crabs together, and do not keep them too long.

- Like fish, crabs absorb oxygen from water through their gills, so change the seawater every 10 minutes to avoid asphyxiation.

- Keep your bucket cool and out of direct sun.

- Pick your crab up correctly by holding it either side of its shell, or from behind with one finger on top of the shell and the other underneath; this way neither crab nor human get hurt in the process.

- Crabs can be… well… crabby; males can be particularly aggressive, so separate any that are fighting.

- At the end of the day, return crabs to the water unharmed, and take your equipment and rubbish home.

- Look after yourself too; in the excitement of the catch, it is easy to slip or fall into the water. Young children should wear lifejackets or buoyancy aids when close to water.

Crabbing together as a family is one of the joys of summer holidays.

How to hold a crab.

Crabbing at Tobermory on the Isle of Mull.

The title of king of the crustaceans must go to the **lobster**, famed not only as a luxury seafood, but also as a fascinating creature. There are half a dozen species found in waters around our islands, and the biggest and most sought after is the **common** or **European lobster**. They are still widespread and fairly common around the British Isles, despite some local overfishing.

When alive, the common lobster is a regal bluey-purple with a paler underside, although they can sometimes verge on black; they only become bright red when cooked. Like other decapods, they have ten legs, and their front two are well-developed claws, quite capable of inflicting serious damage to fingers.

The claws are asymmetric, with the bigger claw used for crushing and holding prey, and the other used for cutting. Common lobsters can occasionally be found in the inter-tidal zone (that part of the coastline between high and low water) but they generally prefer deeper water down to 50m (164ft). They favour a rocky seabed which provides crevices in which to hide, as the smaller lobsters in particular are prey to large fish such as cod, bass and rays. They emerge to feed at night, scouring the seabed for anything that is edible, including marine worms, starfish, other crustaceans and dead or rotting fish.

Lobsters are believed to become fertile around five years old, which is why careful management is important to allow them to reach sexual maturity. Breeding takes place at any time. The female carries her fertilised eggs for up to a year before they hatch. The larvae become free-swimming zooplankton for the first stage of their lives, eventually taking an adult form and settling on the seabed. Although females will carry around 100,000 eggs, less than one percent make it to adulthood.

The common or European lobster (Homarus gammarus) grows to 40cm (16in) long, or more. They have a long, segmented body, protracted antennae and prominent eyes.

Lobster pots are baited with fish and lowered into the water and their position marked with a buoy. The crustaceans climb through a narrow opening but cannot retreat, although traps usually have an escape hatch to allow small lobsters, crabs and fish to escape. Some fishermen maintain that rotting fish is best for lobsters, and that fresh bait attracts crabs, other fish and conger eels. The traps need to be left at least overnight, as lobsters are nocturnal feeders.

Due to commercial pressure, few lobsters grow to their maximum size. It is often claimed that lobsters never die of old age. Whilst this is not strictly true, they are capable of living very long lives; males are typically thought to live to around 30 years, and double this for females.

The **spiny lobster** is also known in Britain as the crawfish or rock lobster, and is mainly confined to the west coast of Scotland, south-west England and Wales, and the west coast of Ireland. They prefer rocky, exposed coastlines, and live in a wide depth range down to 400m (1,312ft). They have a stout, heavily armoured body, usually orange with a white belly, and numerous sharp spines along the shell or carapace, two long antennae and small hook-like claws.

Spiny lobsters were intensively fished in the late 1960s and early 1970s, which led to local extinctions, especially in south-west Britain. It took until 2014 before there were significant signs of their recovery. Like most crustaceans, these scavengers stay hidden during daytime and feed at night.

The spiny lobster (Palinurus elephas) grows to 40cm (16in), but they lack obvious claws. They become noisy during the breeding season to attract a mate, and rub their antennae against their shell to make a creaking noise. They are classified as vulnerable.

CHAPTER 9

THE CARNIVORES

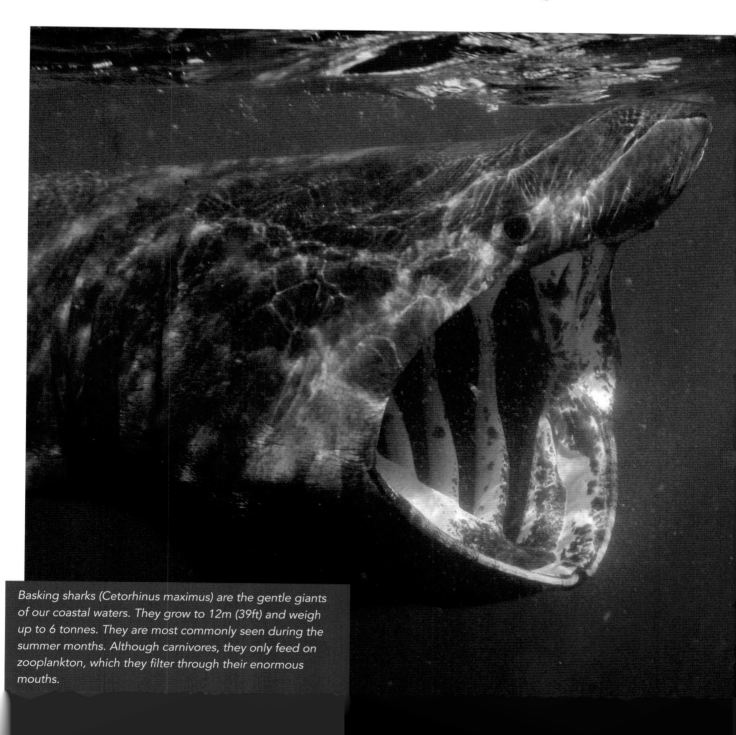

Basking sharks (Cetorhinus maximus) are the gentle giants of our coastal waters. They grow to 12m (39ft) and weigh up to 6 tonnes. They are most commonly seen during the summer months. Although carnivores, they only feed on zooplankton, which they filter through their enormous mouths.

In August 2020, Rupert Kirkwood – an experienced sea kayaker – was watching the abundant sealife off the coast of Plymouth in Devon. His attention was drawn to turmoil in the water – a sure sign that large predatory fish were forcing their prey to the surface in a feeding frenzy. It was not long before a huge silver-blue fish broke clear of the water in pursuit of its prey. It was an Atlantic bluefin tuna – the largest of the magnificent tuna species. These powerful, streamlined predators grow to 4.6m (15ft) and weigh up to 1,000kg (2,200lbs). That day, Kirkwood watched hundreds of bluefin tuna, together with whales and dolphins, feeding on sand eels and herring.

Bluefin are an endangered species, and it is rare to see them in such large numbers so close to the shore in British waters. These spectacular fish were once common around Britain during the 1920s and 30s, but all but disappeared by the 1960s. Now, large shoals of tuna have been reported in the North Sea, the Celtic Sea, and the western English Channel, and this change is a testament both to the improving quality of the waters around our coastline, and the gradual conservation of fish stocks.

FISH
BONY FISH

Bluefin tuna are apex predators at the very top of the food pyramid, alongside sharks and marine mammals such as dolphins and seals. These voracious, fast-swimming, warm-blooded fish eat their own bodyweight every seven to ten days. Because only ten percent of energy is passed up the food chain to higher trophic levels, big hunters like tuna can only survive if fish stocks lower down the food chain are healthy.

Overfishing around the coast of the British Isles has seriously depleted some species, but careful management and conservation should see fish stocks improve in the future. Fortunately, we are blessed with a wide diversity of fish in our coastal waters, with more than 300 species ranging from the tiny, small-headed clingfish, no more than 4cm (1.6in) long, to the magnificent basking shark. Each of these species has adapted to its own ecological niche, which minimises competition for food and maximises their chance of prospering.

The small-headed clingfish (Apletodon dentatus) is one of the smallest marine fish found around our coastline. It inhabits shallow, rocky coastal waters, and is well camouflaged to hide around seagrass and brown algae.

You can tell a lot about how a fish lives and feeds by simply looking at its body shape. Flatfish such as plaice, flounders, turbot and rays are obvious examples, for they have evolved to live and feed on the seabed; the common blenny has a vertically-flattened body which allows it to hide in vertical crevasses in the rocks or amongst seagrasses and seaweeds; fast predators like bass and sharks are streamlined to reduce resistance; the flattened shape of the angler fish is ideal for resting on the bottom and not casting a shadow; full bodied open-water fishes like cod and pollack are a general-purpose shape, making them suited to a wide range of habitats.

Some fishes have adapted sophisticated camouflage to hide from predators. The mottled pattern on a flounder mimics its habitat of sand and gravel; the blue striping on the back of the mackerel and light underside makes it difficult for predators to see, whether looking up towards the sky or down towards the bottom.

The short snouted seahorse (Hippocampus hippocampus).

The silvery herring is hard to see as an individual in a fast-moving shoal of hundreds of fishes; others, such as the John Dory, have a large dark spot on their body looking for all the world like the eye of a much larger fish.

One of the most well-camouflaged fish is the **seahorse**. There are two species in British waters and, perhaps surprisingly, the largest seahorse ever caught anywhere in the world was accidently brought up in a fishing net in Poole harbour in 2015; the long snouted seahorse was 34cm (13.4in) long. Both British species live in shallow coastal waters, with breeding populations along the south coast of England and in the Thames estuary; they are classified as vulnerable and are protected.

Seahorses are poor swimmers, and although they beat their fins very quickly – up to 50 times a second – they move very little, relying instead on their prehensile tails to cling to seaweed and seagrass to stop being swept away.

The long snouted seahorse (Hippocampus guttulatus).

The colourful tompot blenny (Parablennius gattorugine) grows to about 25cm (10in) and lives on rocky shorelines along the south and west coasts of England and Wales. The single branched tentacle above each eye and body colouring offers excellent camouflage, and the narrow body helps it to hide in seaweed and crevices.

The cuckoo wrasse (Labrus mixus) is one of the most colourful fish found in British waters, and rarely exceeds 30cm (12in). Fish develop these distinctive colour patterns for many reasons: bright colours might help with camouflage; they can help fish ambush prey or perhaps protect against predators; and they can also help fish to recognise their own species.

The extraordinary red gurnard (Chelidonichthys cuculus) grows to 40cm (16in). This unusual fish has an armoured head and spines giving it protection. They use their feelers under their head to 'walk' along the seabed when feeding by feeling for worms and crustaceans, and they use their large pectoral fins to effectively 'fly' through the water as they swim.

The sea bass (Dicentrarchus labrax) is a fast and ferocious predator, with a streamlined body, and a large head and mouth. Their distinctive front fin has 8-9 sharp spiny rays. Bass are commonly around 50cm (20in) long, but the largest recorded fish was 103cm (3.4ft). Numbers have declined dramatically, and limits are often imposed on fishing.

The full-bodied Atlantic cod (Gadus morhua) is a widespread mid-water fish, growing to 120cm (3.9ft) and easily recognised by the distinctive barbel under their lower lip. Like many species, young cod live inshore and form large, loose shoals for protection. It is believed that once they exceed about 10kg (22lb), they become solitary and move into deeper water.

The John Dory or St Peter's fish (Zeus faber) has a flat, almost circular, body. They grow to 60cm (24in) at depths of 400m (1,312ft) off southern England and Ireland in summer. They feed on schooling fish, shrimp, squid and cuttlefish. The prominent black spot is said to be the thumb print of St Peter. It acts as a deterrent to predators and confuses its prey.

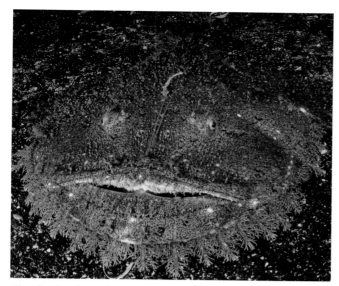

The flat body and camouflage of the angler fish (Lophius piscatorius) helps it to hide on the seabed from predators and prey. It grows to 2m (6.5ft) and is common in the west and south-west. The fleshy growth on its head attracts inquisitive prey, which is quickly despatched with its large mouth and sharp teeth. It is sold commercially as monkfish.

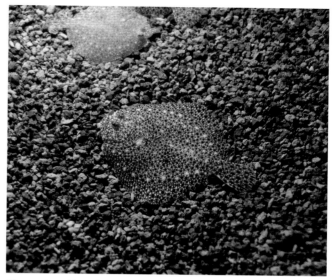

The turbot (Scophthalmus maximus) is found around the British islands, but most common in the south on a sandy or gravel seabed to a depth of 80m (262ft). It grows up to 80cm (31in), occasionally bigger. Its body is nearly circular, about one and a half times as long as it is wide. It can change colour to match its background, and it is a valuable commercial species.

All the fishes described so far have been **bony fish**, with calcified skeletons and bony coverings over their gills.

Shoaling offers protection from predators, with many eyes on the look-out for potential danger. When a shoal is attacked, the fishes co-ordinate sudden movements to confuse a predator. Sometimes a shoal will 'explode', and escape in all directions, but they need to re-group quickly before individual fish are picked off. Experiments have shown that individual fish are less successful when migrating than if they are in a shoal.

CARTILAGINOUS FISH

The other major group is cartilaginous fishes which have a skeleton of cartilage rather than bone, and gill slits which are open to the ocean, rather than covered as in bony fishes. (Our ears, nose and the discs in our backbone are also made of cartilage.) Some of these fish, such as rays and skates, and some sharks, have slits or spiracles on top of their heads, which allows them to rest on the sea bottom and breathe through the top of their heads.

There is fossil evidence that the earliest cartilaginous sharks evolved about 400 million years ago, and rays and skates about 150 million years (around the time of the dinosaurs). Bony fish also date back more than 400 million years, so these two groups of fishes have evolved quite independently for a very long time. Eventually mammals (including humans) evolved from the bony fish side of the evolutionary family.

It was thought that cartilaginous fish were more primitive, but sharks have changed very little over hundreds of millions of years, which suggests they are well adapted to their habitat. Certainly, having a cartilaginous skeleton has several advantages: cartilage is strong and dense, but lighter than bone, which helps a fish conserve energy when swimming at fast speeds; cartilage is also more flexible, which has particular advantages for eels, allowing them to manoeuvre into crevasses and cracks with their pliable body.

There are two species of **eel** found around the British islands, and they both live extraordinary lives. The larger of the two – the **conger** – can sometimes be seen when snorkling over wrecks or rocky reefs (although they prefer to hide during the day). They grow to 1.8m (6ft) or more, but smaller fish can be found under rocks at low water. Conger eels are voracious nightime predators and hunt any fish species, but they will also happily scavenge on dead or rotting carcasses. Congers are extremely fast growing, and one 1.4kg (3lb) specimen in Southport Aquarium grew to 31.3kg (69lb) in just four years; another of a similar size reached 41kg (90lb) after five years.

The conger eel (Conger conger) has a long, powerful snake-like body with smooth, scale-less skin; it is usually grey-blue or grey-black, with a pale belly. They are found mainly around reefs and wrecks, where they wait for their prey. They are widespread along the southern and western coasts of Britain, and around Ireland. The dorsal, tail and anal fins are fused to make a complete fringe along their body. They mature after 5 to 15 years when they migrate to deep water in mid-Atlantic to spawn; they breed only once and die immediately afterwards. The larvae drift as zooplankton until they reach shallower waters on the continental shelf, where they develop into adults.

The smaller relation is the **silver** or **common eel**, and the adults spend most of their lives in freshwater rivers or brackish estuaries. They were once abundant throughout the British Isles and the continent, but numbers have dramatically declined. They feed on fish, worms and crustaceans when in the sea, and on smaller fish, frogs, insect larvae and dead or rotting carcasses when in freshwater.

Parasitic sea lampreys (Petromyzon marinus) firmly attached to an unhappy looking trout.

The common eel (Anguilla anguilla) grows to 1.5m (4.9ft). They are now classified as critically endangered. The adults are a yellowish-golden colour, but turn silvery-blue before they begin their migration to spawning grounds in the Sargasso Sea in the mid-Atlantic. The larvae take about 300 days to drift back to Europe, where they gradually metamorphose into a transparent larval stage called a 'glass eel'. They then enter estuaries and most migrate upstream, where the glass eels grow into elvers, which are miniature versions of the adult eels. © David Pérez (DPC), Wikipedia Commons, Licence cc-by-sa-4.0.

The **sea lamprey** is a primitive, parasitic eel-like fish which lacks jaws, but has a toothed sucker used to latch onto live fish. It grows to 45cm (18in) or sometimes larger. Once fully grown, they clamp themselves to the side of a suitable fish with their flat circular mouth and use their sharp teeth and abrasive tongue to chew through the skin of the host fish and feed on their blood. Fossil lampreys looking very similar to modern species date back 360 million years.

Another primitive jawless relative is the **hagfish**. These are devoted decomposers, scavenging on dead and dying fish. It has been estimated that the discards from trawling for langoustine (scampi) provides up to a third of the food supply for hagfish and other scavengers. Their eyes lack lenses and cannot focus, so they can only distinguish between light and dark and locate their food by smell, assisted by the barbules around their mouths.

Rays and **skates** are bottom-dwelling cartilaginous fishes closely related to sharks but have independently developed a lifestyle similar to flatfish. These fishes have two small breathing holes (spiracles) near their eyes, which allows them to breathe when buried in the sand. There are five species commonly found in British waters. The most common is the **thornback ray**, which grows to 90cm (3ft) and found mainly in the south and west.

The thornback ray (Raja clavata) or thornback skate is found down to depths of 60m (197ft), and feed on crabs, shrimps and small fish.

Some rays have developed powerful attack and defensive mechanisms. The **common stingray**, for example, can inflict painful wounds with its barbed tail. They favour the mouths of rivers and muddy sand. The **marbled electric ray** has an unmistakably rounded shape and is capable of delivering a powerful shock up to 240 volts. They are found in the west and south-west but are scarce. Neither fish should be considered aggressive, but if you stand on one in the water, or catch one when fishing, then you can expect them to retaliate in the only way they know.

The common stingray (Dasyatis pastinaca) grows to 1.2m (4ft) and lives in water no deeper than 60m (200ft) along the south coast of England, but is generally scarce. Though not aggressive, they can inflict a painful wound with a serrated, venomous tail spine. The female gives birth to live young.

The marbled electric ray (Torpedo marmorata) is widespread but rare, and inhabits rocky reefs, seagrass beds, and sandy and muddy flats. This slow-moving predator feeds almost exclusively on small fish, which it ambushes and subdues with bursts of electric current.

The other major group of cartilaginous fish in the British Isles are **sharks** and their close relatives. They do not have scales, but a rough skin with the texture of sandpaper. Perhaps surprisingly, there are 21 resident species of shark found in British waters (including dogfish and smooth hound), with another dozen or so summer visitors.

The most common is the **dogfish** – variously called the lesser-spotted dogfish, the small-spotted dogfish, the small-spotted catshark and the rockfish. They are common and found all around our coastline, except for the east coast.

Most fish reproduce by releasing hundreds of thousands of eggs into the sea, which are mostly eaten by predators or do not hatch for various reasons. The vast numbers released guarantees that a few will mature. Dogfish, however, have a different approach and lay very few eggs, each in a protective case which attaches to rocks or vegetation, where they protect the developing embryo. A female will lay two eggs every five or six days during the breeding season, which runs from November to July. The young hatch after five to ten months, depending on the water temperature. The young dogfish are already 10cm (3.5in) long and able to feed immediately on dead or small prey. After hatching, the empty egg cases are often washed up onshore and are known as 'mermaid's purses'.

The dogfish (Scyliorhinus canicula) grows to 1m (3.3ft) and has a distinctly mottled skin. They prefer shallow waters down to 110m (360ft), where they stay close to the seabed. They are unfussy scavengers and will eat most things, including worms, small fish, prawns, crustaceans, as well as rotting food.

A mermaid's purse washed up on the beach.

The basking shark (Cetorhinus maximus) is classified as an endagered species and offically protected in UK waters.

The **bigger smooth hound** (*Mustelus mustelus*) and **starry smooth hound** (Mustelus asterias) are shallow water shark species which grow to 1.2m (3.9ft). They favour sand and shingle and rarely venture deeper than 100m (330ft). In the past, they were found mainly in the south and west of the British Isles, but are now regularly caught further north. There are two main reasons for this: they are not a commercial species, and they are caught by most sports fishermen on a strict catch-and-release basis. The reward is a thriving species and bigger individual fish.

The big sharks include the **blue**, the **mako** (I*surus oxyrinchus*), occasionally the **thresher**, and the portly filter-feeding **basking shark**. They are all migratory fish and visit our waters as they follow their food north during the summer.

The blue shark (Prionace glauca) is an open ocean (pelagic) species seen in British waters during the summer. They take a clockwise route through the Atlantic, following the Gulf Stream from the Caribbean to European waters in the summer, and returning on a southern route. Blues feed mainly on small fish and squid, although they have been known to take seabirds and other small sharks. The largest blue shark caught in UK waters weighed 116kg (256lbs) and was over 2.7m (9ft) long.

The thresher shark (Alopias vulpinus) has a tail almost as long as its body. When hunting, they use their tail to herd and stun their prey by whipping up the water. They can grow to 4.5m (15ft) and weigh up to 340kg (750lbs).

As our coastal waters become warmer due to climatic change, more exotic species are following their food source north. In the future, we might expect to see the great hammerhead, the bigeye thresher and the ferocious oceanic whitetip shark (which has a menacing reputation for feeding on shipwreck survivors). Some fishermen in the south of England have even claimed to have seen the great white, but to date there have been no independently confirmed sightings, and they are often confused with porbeagles. Swimmers should not be worried as there have been no recorded shark attacks in British waters.

MARINE MAMMALS (CETACEANS)

In December 2019, conservationists announced the year had broken records for sightings of marine mammals. A pod of bottlenose dolphins had migrated from Scotland to East Yorkshire – the furthest south they had been officially recorded, and minke whales were seen feeding off Staithes in North Yorkshire. In Scotland, the Hebridean Whale Trail was launched, which identifies 33 places along the west coast of mainland Scotland and its islands to see whales, dolphins and porpoises – and you do not even need to book a boat trip, because these are all viewing points along the shoreline; you could be lucky and also see basking sharks, seals, and other marine life.

All these sharks and marine mammals are at the top of the food chain – the lions and tigers of the oceanic world. The hot summer of 2019 resulted in our coastal waters being particularly warm, creating perfect conditions for phytoplankton to flourish; this in turn attracted more fish to our waters, and with them came the apex predators.

There are few sights at sea more uplifting than watching dolphins and whales in their natural habitat, and this is now becoming a more common sight in British waters. Feeding habits vary between species; some are partial to squid and cuttlefish, whilst other prefer mid-water fish such as cod, herring and sardines.

DOLPHINS

The **white-beaked dolphin** is the most common in northern Britain and can be identified by indistinct white patches on its sides – not to be mistaken with the Atlantic **white-sided dolphin**, which has conspicuous long, white, oval patches on the flanks behind the dorsal fin. The **bottlenose dolphin** is probably the most familiar dolphin, but not easy to identify as it has no characteristic markings, and its distinctive stubby beak-shaped nose is not easy to spot when they are swimming.

The white-beaked dolphin (Lagenorhynchus albirostris) are oddly named as most lack this feature. They are predominantly black with grey and white flanks. They eat a wide range of food, and grow to about 3m (10ft) and 350kg (770lbs).

The Atlantic white-sided dolphin (Lagenorhynchus acutus) are highly sociable, often coming together to form large pods containing hundreds or thousands of dolphins. They have distinctive markings, making them easy to identify. They grow to 2.8m (9.1ft) and 230kg (500lbs).

The bottlenose dolphin (Tursiops truncatus) grows to 4m (13ft), with small resident populations in the Moray Firth, Cardigan Bay, south-west England and western Ireland. They feed on a wide range of fish, including salmon, sea trout, mullet, herring, sprat, mackerel, squid and cuttlefish.

Sightings of the short-beaked common dolphin (Delphinus delphis) have boomed in recent years. They are most likely seen from Cornwall to the Inner Hebrides and travel in large groups of hundreds of individuals. They grow to 2.7m (8.9ft) and 140kg (310lb).

The Risso's dolphin (Grampus griseus) has a distinctive blunt head, no beak, and scars on its body from social interactions or attacks from squid or octopus. They are highly sociable, but boat shy. They primarily eat squid, and grow to 4m (13ft) and 500kg (1,100lbs).

Long-finned pilot whale (Globicephala melas) is the largest species of dolphin and grows to 6.7m (22 ft) and 2,300kg (5,070lbs). Pilot whales get their name from the belief that there was a lead individual or 'pilot' in each group. Pilot whales are more likely to beach themselves than any other cetacean.

Related to dolphins are **orcas** or killer whales which, despite their common name, are not whales. (Orcas were originally called 'whale killers' by ancient sailors, who watched them attack large whales, and their name was flipped to the more memorable 'killer whale').

Their feeding habits are very different, as they are the ultimate apex predator. No other animals (except for humans) hunt orcas, and they feed on seabirds, squid, octopus, sea turtles, sharks, rays and fish, as well as other marine mammals. There are few animals in the ocean capable of taking on a large whale or a great white shark and winning. Orcas will sometimes pursue their prey into shallow water, but despite their aggression, they are not a danger to humans.

These highly intelligent mammals are very social and live together in groups called pods. Members of resident pods are generally less aggressive and seem to prefer feeding on fish, whereas transient or migratory pods are more bellicose and feed like wolf packs, hunting efficiently by working together. We have both groups in British waters. Migratory orcas arrive in northern Scotland and the Scottish islands in early summer. There is also a resident group in Scotland known as the 'west coast community'. In 2016, the group lost a key female called Lulu, who was found washed up on the Isle of Tiree after becoming tangled in a fishing line. This left just four males and four females; however, there have been no births for more than 25 years, and nobody understands why. This most likely leaves Scotland's only resident pod facing extinction.

WHALES

There are seven species of true whales found around our coastline – either filter-feeding baleen whales, or toothed whales. Baleen whales are generally larger, and use their baleen plates to filter food – usually krill and juvenile fish. All baleen whales have two blowholes, and some species are well adapted with a thick layer of fat or blubber to keep warm in cold water, or when diving to great depths.

Toothed whales are more closely related to dolphins and porpoises, and are opportunistic feeders, generally seeking squid and fish. The sperm whale is the only known predator of the giant squid. Toothed whales are intelligent animals, and the sperm whale has the biggest brain mass of any animal, except for humans.

The orca (Orcinus orca) is widespread throughout the world's oceans, but only found on the west and northern coasts of Scotland and offshore islands. Males grow to 9m (29.5ft) and weigh over 6 tonnes. They can live to 100 years, and are the only cetacean known to prey on other marine mammals; they have even been known to attack a lame blue whale.

BALEEN WHALES

The fin whale (Balaenoptera physalus) is also known as the common rorqual. This baleen whale feeds by filtering plankton. Like all large whales, they were heavily hunted, and are classified as endangered. They are the second largest animal in the world, and typically grow to 25m (82ft) and weigh 74 tonnes.

The humpback whale (Megaptera novaeangliae) is another filter feeder and known for spectacular breaching. They sometimes feed as a group, and use a technique called 'bubble netting', where a pod creates a curtain of bubbles around a shoal of krill or small fish, trapping them inside. They grow to 16m (52ft) and weigh around 30 tonnes.

Numbers of the sei whale (Balaenoptera borealis) fell dramatically due to commercial whaling during the late nineteenth and twentieth centuries, when over 250,000 whales were killed. The sei whale is now internationally protected and its worldwide population is less than a third of its pre-whaling population. They are among the fastest whales, and grow to 18.6m (61ft) and 22 tonnes. They are well-documented accounts that sei whales are prone to attack by orcas.

The minke whale (Balaenoptera acutorostrata) is one of the smallest baleen whales, and grows to 8m (26ft) and around 10 tonnes. They undertake seasonal migration to the poles in spring, and back to the tropics in autumn and winter.

TOOTHED WHALES

The sperm whale (Physeter macrocephalus) is the biggest of the toothed whales, and grows to 20.7m (68ft) long and up to 80 tonnes. Sperm whales were prime targets for whaling, as sperm oil was used in the past for lamps, lubricants and candles. The epic struggle between whale and whaler was depicted in the book Moby Dick by Herman Melville.

The northern bottlenose whale (Hyperoodon ampullatus) was hunted heavily by Norwegian and British whalers during the nineteenth and early twentieth centuries. They are easily recognised by 'bulbous melons' on their large foreheads. They grow to 11m (36ft) and 7.5 tonnes.

The Cuvier's beaked whale (Ziphius cavirostris) or goose-beaked whale feed on squid and deep-sea fish. They are one of the most common of the beaked whales and can be seen in northern waters around the Scottish islands. They grow to 7m (23ft), weighing up to 3.5 tonnes.

All cetaceans have legal protection throughout European waters, but they are sometimes caught and die during commercial fishing. When tangled in nets, dolphins and porpoises cannot surface to breathe and consequently suffocate; this is called 'incidental by-catch'. Attempts are made to reduce accidental deaths, and fishing vessels over 12m (39ft) using bottom set gill or entangling nets must have acoustic deterrent devices fitted, designed to discourage cetaceans from approaching the fishing gear. Even so, in 2019 alone, more than 1,100 harbour porpoises died in fishing nets in UK waters, according to the charity Whale and Dolphin Conservation.

Shark, Whale & Dolphin Watching

Wherever you might be around the coast, there are likely to be boat trips available in season to watch these apex predators in their natural habitat.

Species	Region	Best time to observe
Blue shark (*Prionace glauca*)	South-west coast, normally 16km (10mi) or more offshore	June to October
Thresher shark (*Alopias vulpinus*)	All UK waters	Summer
Basking shark (*Cetorhinus maximus*)	All UK waters, but most frequently SW England, Wales, Isle of Man and the west coast of Scotland	May to September
Porbeagle shark (*Lamna nasus*)	All UK waters	June to October
Risso's dolphin (*Grampus griseus*)	Wales and east Scotland	All year
Atlantic white-sided dolphin (*Lagenorhynchus acutus*)	Northern isles, Outer Hebrides, north Scotland, occasionally in northern North Sea	All year, but especially July to September
Common dolphin (*Delphinus delphis*)	All UK waters, but most common in south and west, and offshore	All year
Bottlenose dolphin (*Tursiops truncatus*)	All UK waters, but common in the Moray Firth (Scotland), Cardigan Bay and the Llyn Peninsula (Wales) and Devon and Cornwall (England)	All year
White-beaked dolphin (*Lagenorhynchus albirostris*)	Scotland, north-east England, Lyme Bay (southern England)	All year, best in summer in north-east England
Harbour porpoise (*Phocoena phocoena*)	All UK waters	All year
Orca / Killer whale (*Orcinus orca*)	North and west Scotland, especially Shetland	All year
Fin whale (*Balaenoptera physalus*)	North-east England, Celtic Deep	Summer (offshore)
Humpback whale (*Megaptera novaeangliae*)	UK wide, most common off the Shetland Isles and Hebrides	Sporadic sightings all year
Sei whale (*Balaenoptera borealis*)	Rare but sighted in Firth of Forth. (Scotland)	Spring and autumn during migration
Minke whales (*Balaenoptera acutorostrata*)	UK wide	All year
Sperm whale (*Physeter macrocephalus*)	Rare, mainly north and west coasts of Scotland	July to December
Northern bottlenose whale (*Hyperoodon ampullatus*)	Shetland Isles, occasionally northern Scotland	Mainly August
Cuvier's beaked whale (*Ziphius cavirostris*)	Outer Hebrides and south-east of the Shetlands, but only rare sightings	All year

SEALS

Of all the marine mammals you are most likely to see around our coastline, seals come top of the list. Britain is home to about 40 percent of the world's population of **grey seals**, and they are the country's biggest native carnivore. At the beginning of the twentieth century, hunting brought their numbers down to only 500 individuals, but conservation has increased their population, and there are now thought to more than 120,000.

A grey seal (Halichoerus grypus) and a pup enjoying the sunshine. They prefer uninhabited islands and often return to the same beach every year to breed. Seals give birth to a single pup, which the mother sniffs to learn its scent. Weaning takes about three weeks, by which time the pup has moulted its white fur.

Early September marks the beginning of the pupping season, and it generally lasts through to January. This is the ideal time to observe them when mothers and pups bask on beaches, sandbanks and rocky outcrops. A typical pup weighs 15kg (33lbs) and suckles its mother's milk, which is high in fat; as a result, the pups can triple their birth weight in the first three weeks. The Norfolk coast is a popular breeding ground for the marine mammals, along with several sites in Cornwall and Scotland.

The other resident seal species is the **harbour** or **common seal**, and there are thought to be 50,000 individuals in British waters. The common seal is smaller than the grey, and has a shorter head, a more concave forehead, and V-shaped nostrils. They are generally grey with dark spots, but colours vary from blonde to black. They give birth in the summer.

The smaller common seal (Phoca vitulina) is found around the coasts of Scotland, Northern Ireland, north-east and eastern England. Like grey seals, they feed on fish, squid, whelks, crabs and mussels.

Seals are protected by an act of parliament, but they are still legally culled in Britain. Several hundred are killed every year in Scotland by owners of fish farms to protect their stocks. Plastic pollution, entanglement in discarded fishing nets and disturbance by tourists are also threats. Pups are inquisitive by nature and vulnerable as they begin to explore their world and will often eat plastic objects; they have been known to starve to death after filling their stomachs with plastic bags.

Viewing Seals

If you get an opportunity to view seals:

- Do not get closer than 20m (66ft), or the length of two buses.

- Avoid getting between a mother and her pup, or any seal and their escape route to the sea.

- Keep noise to a minimum and your dog on a lead.

LAND MAMMALS

There are a couple of mammals which frequent our coastal waters, and which you might see if you are patient. The European **otter** hunts in wetlands and along rivers and beaches, but they are elusive. They are lithe swimmers and feed mainly on fish (particularly eels, salmon and trout), water birds, amphibians and crustaceans. They are well suited to life in the water as they have webbed feet, dense fur to keep them warm, and can close their ears and nose when underwater. The best places to see them are the west coast of Scotland, west Wales, the West Country and East Anglia.

The European otter (Lutra lutra) grows to 80cm (30in) plus its tail, and lives for 10 years in the wild. They are powerful mammals with grey-brown fur and a pale chest and throat; they can be distinguished from the darker mink by their larger size and broader face.

The **mink** is a semi-aquatic, carnivorous mammal related to the otter. They are not native to the British Isles, but an American species which escaped or were released from mink farms. Today, they are widespread throughout mainland Britain, except for mountainous areas in Scotland, Wales and the Lake District. Although mink are not specifically aquatic animals, they feed on fish and other aquatic life, small mammals, birds and eggs. They are smaller and darker than otters. It can be argued that mink have now become part of the British ecosystem and fill the role of native predators such as otters, which have disappeared in some areas because of hunting.

The American mink (Neovison vison) adapts well to an aquatic life, and will scavenge on a wide variety of food found around the coast. They are not easy to see as they prefer to hunt and feed at night. They are significantly smaller than otters, growing to around 50cm (20in) plus their tail.

CHAPTER 10

COASTAL BIRDS

The herring gull (Larus argentatus) is a large, noisy gull which can be found all year around our coasts. They can easily be identified by a red spot under their curved bill. The young are mottled brown.

In 1824, the last pair of nesting avocets was seen in Salthouse in Norfolk. The drainage of coastal wetlands, land reclamation for farming and overly enthusiastic Victorian egg collectors finally drove this pretty little wading bird to national extinction. 120 years later, coastal areas around eastern England were flooded during the Second World War as a defence against invasion, and this threw a lifeline to this distinctive coastal bird. As allied forces prepared for the invasion of mainland Europe, breeding pairs (probably from the Netherlands and Denmark) came the opposite way and recolonised their old habitats. The avocet was finally back in England.

The avocet's increase in numbers since the 1940s is one of the most successful conservation projects in the country. The bird was adopted as the emblem of the RSPB and has come to symbolise the bird protection movement in the UK more than any other species.

HABITATS

With over 17,820km (11,070 miles) of coastline in Great Britain alone, we have a long and varied habitat to suit almost every bird likely to visit our islands. As with fish and other marine animals, birds have adapted to live and prosper within their own ecological niche, and to minimise competition with others. Many coastal birds come close to the top of the food chain as carnivorous predators and scavengers, others are omnivorous, and some are strictly vegetarians.

Many bird species live on or near the coast, but that does not necessarily make them true seabirds. You often find wildfowl such as ducks and geese around estuaries or coastal marshland, and ravens, doves and kestrels frequent sea cliffs. They might live by the sea, but they are not true seabirds, which are generally considered to be species that have specifically adapted to the marine world and rely most of the time on the ocean to survive.

The opposite is also true. Gulls are most definitely seabirds, but they frequently colonise inland waste sites and scavenge in cities. Even though they might spend all their lives scrounging food in Trafalgar Square, they are still considered seabirds.

The coastal environment is a challenging place for any bird to survive, and they have adapted to cope with extreme conditions in many different ways. Seabirds inevitably have a diet high in salt, and they have developed special glands like miniature kidneys next to their eye sockets. These remove excess salt from the bird's bloodstream, where it drips harmlessly down the side of their bills. Their feathers have adapted to keep the birds dry and well-insulated, and most seabirds seem to be indifferent to all but the worst stormy conditions. Most seabirds also have flexible webbed feet, which help them swimming or to gain speed when taking off from water.

THE OPEN OCEAN

Some birds have adapted to life on the open sea, and only ever come ashore to roost. These are the true oceanic birds, and include **petrels**, **fulmars** and **shearwaters** (all related to the albatross), as well as **skuas** and **auks**. These oceanic birds are sometimes known by the term pelagic, which means 'of the open ocean'. Like many long-distance fliers, these birds cannot walk properly on land, which makes them particularly vulnerable to predators when they come ashore. They are in their true element at sea, where they are perfectly adapted to use the winds and air currents to fly and swoop efficiently over vast areas of the ocean, using minimal effort.

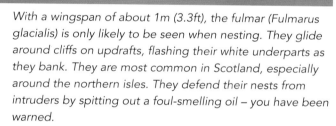

With a wingspan of about 1m (3.3ft), the fulmar (Fulmarus glacialis) is only likely to be seen when nesting. They glide around cliffs on updrafts, flashing their white underparts as they bank. They are most common in Scotland, especially around the northern isles. They defend their nests from intruders by spitting out a foul-smelling oil – you have been warned.

The storm petrel (Hydrobates pelagicus) has a wingspan of less than 40cm (16in), and is no bigger than a song thrush. They flutter over the water rather than glide, flashing their white rump against their black plumage. They often feed in flocks, following in the wake of ships, especially trawlers. They are found all along the west coast of the British Isles.

The great shearwater (Puffinus gravis) is a large seabird with a wingspan up to 1.2m (3.9ft). They are only seen offshore, mainly along the north and west coasts of Scotland and the south-west coast of England. They are most likely to be seen from headlands from July to October, when the winds are onshore.

SEA CLIFFS

All birds need to hatch and raise their young ashore, and coastal cliffs offer seabirds the big advantage of being inaccessible to most predators, including humans. Depending on the geology, these habitats offer a wide variety of nesting sites, and give security for mixed seabird colonies. The **auk** family in particular, is highly adaptable to cliff-top living: this group includes razorbills, guillemots, puffins and little auks. Most of these seabirds have difficulty walking, but they can at least manage to stand upright on cliff ledges.

Guillemots are extremely gregarious birds and are found mostly in northern England and Scotland. When they come ashore to breed, they nest in tightly packed colonies, or 'seabird cities', on perilous ledges, clifftops and rocky outcrops such as Bass Rock. They are a particularly successful seabird, with an estimated 950,000 breeding pairs around our islands.

One of the favourite seabirds must be the **puffin**, often nicknamed the 'sea parrot'. Out of the breeding season they lead a solitary life at sea, bobbing about on the surface like corks. They form long-term relationships, and in spring the mature birds return to

nest, usually to the colony where they were hatched. Their diet consists almost entirely of fish, although they occasionally eat worms, shrimp and other crustaceans. When fishing, the birds swim underwater using their semi-extended wings as paddles, and their feet as rudders. They can dive to 60m (200ft) and stay submerged for up to a minute. The four best places to see these pretty birds is Skomer Island in Pembrokeshire, Bempton Cliffs in North Yorkshire, South Stack Cliffs on Anglesey, and St Kilda off the west coast of Scotland.

The unmistakable puffin (Fratercula arctica) returns with a mouthful of sand eels for its chick. These small birds have a wingspan of about 60cm (2ft), and are most common in the north-east and south-west of the country. Their beak changes colour during the year; in winter, it is a dull grey, but in spring it blossoms into a striking orange, which is thought to help them attract potential mates.

A sole razorbill (Alca torda) on the left, joins a colony of guillemots (Uria aalge) on the Farne Islands, Northumberland. The larger guillemot has a wingspan to 73cm (29in), and both species only come ashore to breed.

The little auk (Alle alle) is the smallest of the auk family with a wingspan no more than 48cm (18in). They fly with fast, whirring wingbeats low over the sea, feeding on plankton and small fish. The bird breeds in the Arctic and is a winter visitor along the east coast of the British Isles, but only in relatively small numbers.

Various **gull** species, **shags** and **gannets** all add to the diversity of cliff-dwellers, and Bass Rock in the Firth of Forth is renowned for its seabird colonies. The two things that you notice more than anything as you approach Bass Rock (especially if you are downwind) are noise of hundreds of thousands of birds and the pungent smell of ammonia from years of accumulated **guano** – the excrement of seabirds – which makes a highly effective fertiliser.

This ancient volcanic plug is home to over 150,000 gannets at the peak of the breeding season. The birds spend most of the year on the rock until October, when they start their long migration south, with many going as far as the west coast of Africa.

The lower ledges of Bass Rock are home to shags, as well as guillemots and razorbills, with seals hauling themselves up on the rocks below. The common shag is the size of a goose, with a long, dark neck and a steep forehead. During the breeding season, adults develop a dark glossy green plumage and a crest on the front of their head. Shags can often be sighted on a rock with their wings proudly held out. These seabirds nest on coastal cliffs, and are rarely seen inland.

A pair of shags (Phalacrocorax aristotelis) on the cliffs of the Farne Islands, Northumberland, where they feed on fish, and occasionally crustaceans and molluscs. Shags have a wingspan up to 1m (3.3ft); they are closely related to the bigger cormorants, but can be distinguished by the pronounced crest on their forehead.

Northern gannets (Morus bassanus) on Bass Rock in the Firth of Forth. They are large birds, with a wingspan of up to 1.8m (6ft). They feed by flying high and looking for fish shoals, then plunge into the sea after their prey, hitting the water at speeds of up to 100km/hr (60mph). Gannets arrive at their breeding colonies from January onwards, and leave in autumn. They can be seen all around the UK.

Gulls do not have the best of reputations. They are scavengers by nature, and just as happy foraging over a municipal landfill site as they are circling coastal cliffs. All six main species of gulls prefer remote and inaccessible locations, and they also nest along sand and shingle shorelines, and in sand dunes. In the last few decades, many birds have moved inland, where they thrive in habitats that we have created. Cities and their rubbish tips now sustain many more gulls than live in their traditional marine habitats, and we only have ourselves to blame.

But things are changing; most gull-edible trash is now composted or incinerated, and if councils empty their waste bins and day-trippers hang on to their fish and chips, then gulls will change their behaviour and move back to their natural habitat.

There is actually no such bird as a 'seagull', although the term is commonly (but incorrectly) used to describe any of the gull family, which includes the herring gull, the lesser black-backed gull, the great black-backed gull, the black-headed gull, the common gull and the kittiwakes.

In England, the chalk cliffs around Flamborough in Yorkshire are now the biggest mainland seabird colony in the British Isles and home to over 400,000 birds, including gannets, puffins and razorbills. Other large seabird colonies can be found at Longhaven Cliffs in Aberdeenshire, Scotland, the Isle of Muck in Co. Antrim, Northern Ireland, and Skomer and Skokholm Islands in Pembrokeshire, Wales.

The black-legged kittiwake (Rissa tridactyla) is a member of the gull family, and spends winter at sea. With a wingspan up 1.1m (3.6ft), they are smaller than the herring gull, but can easily be identified by their distinctive nasal repeated call, and by their white underbelly, pale grey back, and distinctive black wing tips. They nest in clifftop colonies from February until August.

The European herring gull (Larus argentatus) has a wingspan up to 1.5m (5ft); these large, noisy gulls are found throughout the year around our coasts, and scavenge inland, especially in winter. They are not fussy eaters, and are happy to take carrion, offal, seeds, fruits, young birds, eggs, small mammals, insects and fish.

ROCKY SHORELINES

These austere parts of the coastline do not offer much of a habitat for seabirds, and only species such as the **purple sandpiper** and the **turnstone** make a success here. They are able to feed in this challenging habitat by using their sharp, pointed bills to hunt out insects, crustaceans and molluscs from amongst the rocks and seaweed.

The purple sandpiper (Calidris maritima) is a winter visitor to almost any rocky coast in the UK, although they are mainly found in Orkney, Shetland and along the east coast of Scotland and northern England. A couple of pairs in Scotland stay over the summer and nest, but the location is kept secret to protect them from egg thieves and disturbance. These small birds have a wingspan up to 44cm (17in), and feed on insects, spiders, crustaceans and plants.

The turnstone (Arenaria interpres) in winter plumage, is a non-breeding visitor. It grows to 25cm (10in) with a wingspan up to 57cm (22in). They are one of the few birds that flourish on rocky coastlines, and spend their time fluttering over the rocks looking for food. This small bird is capable of lifting rocks as big as itself and is a common winter visitor to our coast. Turnstones can be found all round our coastlines during winter.

SAND & SHINGLE

About one third of the UK coastline is fringed with sand or shingle beaches, and these areas offer another harsh environment with little soil, few plants, practically no freshwater, strong winds and salt spray. Yet these inhospitable sites provide an attractive habitat for terns and gulls to establish nesting colonies, especially if the area is rarely disturbed by humans.

During the summer, five species of **tern** return to nest along our shorelines, the highest numbers being the **common tern** and the closely related **Arctic tern**. All terns have distinctive grey bodies, white bibs, black head-caps, and pointed beaks, although they vary in size from the little tern, with a wingspan of only 50cm (20in), to the common tern which is double that size. The birds nest in well-camouflaged shallow scrapes on sand and shingle beaches, spits or inshore islets, with little or no nest material.

The common tern (Sterna hirundo) can be recognised by its long tail streamers, hence its local name of 'sea swallow'. With a wingspan up to 1m (39in), they frequently hover over the surface before plunging down for a fish. Common terns are often noisy in company and breed in large colonies; they are also found inland. These birds migrate from Africa, South America and south-east Asia, with a typical round trip of more than 30,000km (18,600 miles).

The Arctic tern (Sterna paradisaea) is not too choosy about finding a suitable nesting site, although they prefer remote places. They will attack anyone approaching their nest, often dive-bombing with their sharp bills. These are summer visitors that nest on islands around the north of England and Scotland, with large colonies on the Farne Islands in Northumberland, and the northern isles. Arctic terns have the longest migration of any bird and return to the Antarctic every year for the southern summer.

WETLANDS

Tidal flats, salt marsh and wetlands provide a wide-ranging diet for specialist feeders, and they are important stop-over sites during migration. Most of the species found in wetlands are not true seabirds, but are commonly known as 'waders', and are perfectly adapted to this amorphous zone between water and land. These birds typically have long legs, and their bills are well-adapted to feeding. They can often be seen half-submerged along shorelines, river mouths and estuaries as they feed on a rising tide.

The **avocet** is an elegant wading bird that is fascinating to watch feeding in muddy water, as it forages by touch, sweeping its long, upcurved bill from side to side through water or loose sediment to locate hidden prey. In clear water, they feed by sight, picking their food from the water or mud surface. They are accomplished, buoyant swimmers, and in deeper water they up-end like a duck to reach their food. You sometimes see large flocks comprising several hundred individuals feeding co-operatively on a rising tide on such delicacies as shrimps and worms.

During the breeding season, avocets favour brackish lagoons with sparsely vegetated islands and gently sloping banks, generally close to the coast. In north Norfolk, these conditions are found at many of the managed coastal reserves, where protection – at least from human disturbance – can be provided. In Norfolk, Cley Marshes, Hickling Broad and Holme Dunes are good places to see them. Further south in Suffolk, you can see them at Dingle marshes, and in Essex the birds obligingly nest in front of one of the hides at Blue House Farm in North Fambridge.

The **oystercatcher** is another unmistakable wader, with black and white plumage, beady red eyes, a long orange-red bill, reddish pink legs and a loud alarm call. The birds breed all around the British Isles, and in the last 50 years they have also started breeding inland. Most resident birds spend the winter on estuaries and mudflats, where they are joined along the east coast by birds from Norway.

Nearly all species of oystercatcher are monogamous, and they are highly territorial during the breeding season. Their nests are simple scrapes in the ground, placed in a position with good visibility. Oystercatchers also have a reputation for 'egg dumping' (rather like the cuckoo) and sometimes lay their eggs in the nests of other species, such as gulls.

The diet of the oystercatcher (Haematopus ostralegus) rarely includes oysters, and they mainly feed on bivalves, including cockles and mussels. The birds have two techniques to open difficult prey; those with shorter, blunter bills, specialise in hammering the shell open, whilst those with longer, more pointed bills, prise the shells apart. These medium sized birds have a wingspan up to 86cm (34in).

The unmistakable avocet was adopted as the emblem of the RSPB and has come to symbolise the bird protection movement in the UK more than any other species. The wader's increase in numbers since the 1940s is one of the most successful conservation projects in the country.

The avocet and oystercatcher are just two of about 15 species of wading birds that live around our estuaries and wetlands. Because of our mild winters, many millions more migrate around our coastal waters to create one of the country's great wildlife spectacles. Small waders like the dunlin, several sandpiper species, stints, snipe, redshanks, curlews, and godwits are common visitors to our muddy, shallow coasts. One of the easiest to recognise is the curlew, a large wader with mottled plumage and a distinctive downward-curving bill.

With a wingspan up to 1m (3.3ft), the curlew (Numenius arquata) is the largest wader around the British coastline, and in the same family as sandpipers, snipe and phalaropes. They can be seen in Morecambe Bay, the Solway Firth, the Wash, and estuaries of the rivers Dee, Severn, Humber and Thames. They feed on worms, shellfish and shrimp.

The common snipe (Gallinago gallinago) is a smaller wader with a wingspan of about 46cm (18in). They live in marshes, wet grassland and moorland. They use their long, probing bill to search for insects, worms and crustaceans in the mud, typically swallowing their prey whole.

The five species of **grebe** found in the British Isles have a similar lifestyle, and have all the usual physical features of wading birds. Their legs are set well back on their bodies, and their ankle and toe joints are exceptionally flexible, allowing their webbed feet to be used as both paddles and rudders. They are so well adapted for life in the water that they are really quite ungainly on land, and usually build their nests on floating rafts of vegetation for safety and to save walking. They feed mainly on fish, small eels, crustaceans, molluscs, amphibians and some plants.

The great crested grebe (Podiceps cristatus) is common around the coastline in winter although, like divers, they usually move inland to breed in spring. This elegant bird with ornate head plumes was hunted almost to extinction in this country for its feathers.

Other specialist feeders include **divers** and **marine ducks**, although these species are usually classified as water birds. Three species of divers are in the UK regularly (two of which breed), including the red-throated, the black-throated, and the great northern diver. All divers all have long, slender bodies and dagger-shaped bills. Their long wings are narrow, and like the grebe, their small legs are set well back on their body. This makes them efficient fliers and expert swimmers, but they are clumsy on land and are barely able to walk or even stand. All divers breed on freshwater lakes and pools but move to the coast in winter.

The great northern diver (Gavia immer) is a winter visitor. They use their pointed bill to probe vegetation for food, and hunt for fish and crustaceans by swimming with their head down and their eyes underwater. They are called common loons in North America.

Wetlands are particularly important outside the breeding season; **wildfowl** such as **ducks** and **geese** mainly have a plant diet, so these areas provide essential stop-over sites during their migration. **Swans** are the largest group of wildfowl in the British Isles: the **mute swan** is resident, and the **Bewick's swan** and **whooper swan** visit from autumn to spring, migrating long distances from their summer breeding grounds near the Arctic. Their long necks allow them to take vegetation from the bottom, and mute swans will also graze on molluscs, small fish, frogs and worms. Bewick's and whooper swans can often be seen during the day eating leftover grain and potatoes on fields, before returning to roost in open water.

The mute swan (Cygnus olor) has a wingspan up to 2.4m (7.8ft) and can be seen throughout the year in the UK. The British monarch has claimed ownership of all these birds since the twelfth century because the birds were highly valued as a delicacy at banquets and feasts. Numbers have increased in recent years, due to better protection. They can be identified by their predominantly orange beak with a black base.

Bewick's swan (Cygnus columbianus) is the smallest seen in the UK, with a wingspan up to 1.9m (6.2ft). These birds nest in Siberia and migrate more than 5,000km (3,100 miles) to spend winter in the UK, mainly in East Anglia, the Severn estuary and Lancashire. The orange markings on its bill are less extensive than the whooper swan.

With a wingspan of up to 2.3m (7.5ft), the whooper swan (Cygnus cygnus) is the largest of our three swans. They are mainly winter visitors from breeding sites in Iceland, although a small number of pairs nest in Scotland. They can be seen in estuaries and wetlands in Northern Ireland, Scotland, northern England and parts of East Anglia.

FEEDING

Most coastal birds find their food by sight. This led to a belief that they had a poor sense of smell, but petrels have been found to have a well-developed sense which allows them to find food in the open ocean. The bills of coastal birds will tell you a lot about their lifestyle. Offshore fish-eating seabirds such as gannets, terns and guillemots have dagger-like, pointed bills – very useful for diving into the sea for fish.

The herring gull is an opportunistic feeder, and its hook-like bill is just as useful for digging for earthworms from the soil, as it is for scavenging from a waste bin. Razorbills and shags also have a curve to the end of their bill, which helps them hold on to a slippery catch. The puffin has a short, stumpy bill which allows them to grab up to a dozen small fish at a time, without having to regurgitate swallowed fish to feed their young.

Waders, such as the curlew and snipe have long bills, which are perfectly adapted for digging for worms on mudflats. In many species, the upper part of the bill can flex upwards, allowing them to open their bill to grab food even when buried deep in the mud. Most species have sensitive nerve endings at the end of their bills, which enables them to detect food hidden in mud or soft sand. Sandpipers and turnstones, on the other hand, have short, stubby beaks which allow them to dig out food from rock crevices and around seaweed.

RAPTORS

At the top of the avian food chain come the raptors – birds of prey – with their effortless aerobatics and ruthless hunting instinct, they are the undisputed masters of the skies. There are 15 raptor species in the British islands, and they fall into three basic groups: **eagles**, **hawks**, and **falcons**. Of these, only a few can be considered true coastal birds.

Birds of prey have several things in common, apart from the fact that they have been vulnerable to persecution and egg collecting. They all eat flesh, but some specialists such as the osprey prefer fish. The larger raptors are not too fussy, and are more likely to scavenge, eating everything, including feathers, down and bone. Their stomachs are so acidic that they dissolve the bones of their prey, and any indigestible material is regurgitated as pellets.

Female raptors definitely rule the roost, and they are always bigger than their mates. During the spring nesting season, the cock will feed both the female and the young, until she is able to start hunting again – and when she does, her bigger size allows her to catch even bigger prey. Unlike other bird species, the young hatch at intervals so, if there is a shortage of food, the oldest and largest chick is the one that survives.

The smallest of the coastal raptors is the **marsh harrier**, distinguishable from inland harriers by their heavier build, broader wings and lack of white on the rump. These birds were once thought to frequent only areas of marshland and reed beds, but they are now just as likely to be found hunting and nesting on arable land, where they feed on small birds and mammals. The future of these birds looks more assured than at any time in the last 100 years but, because of population declines in the past, they remain on the amber protection list.

The European marsh harrier (Circus aeruginosus) has a wingspan up to 1.3m (4.3ft), and is the largest of three species of harrier in the UK. They nest in large reed beds and feed on frogs, small mammals and birds, such as moorhen and coot. During the breeding season, the males perform extraordinary courtship displays by wheeling in the air, then diving towards the ground while performing a series of tumbles.

The **osprey** is a remarkably well adaptable bird and one of the most widely distributed, with the same species found as far apart as Alaska and Australia, and from Ecuador to Ethiopia. Even so, the bird was a victim of persecution and Victorian egg collectors; they became extinct in the British Isles as breeding pairs in 1840, then disappeared completely by 1906. In 1954, migrating birds from Scandinavia began to recolonise naturally in Scotland, although early progress was slow and sporadic due to the contamination of their food from pesticides and the activities of illegal egg collectors.

Today, ospreys are protected, and their nests carefully guarded. You are most likely to see them in Scotland, although they are now summer visitors to England and Wales. They are the only British avian raptor with a taste for fish, and with a large wingspan they are an impressive sight when they hit the water hard and emerge with a fish skewered in their talons. They have a conservation status of 'amber'.

The osprey (Pandion haliaetus) has a wingspan of 1.5m (5ft). They overwinter in west Africa and return to the UK each spring to refurbish their large stick nests and breed. There are now around 300 nesting pairs in Britain.

The largest bird of prey in the British Isles – bigger even than the golden eagle – is the **sea eagle** or **white-tailed eagle**, sometimes called the flying barn door because of its huge 2.4m (8ft) wingspan. In flight, its long, broad wings have distinctive 'fingered' ends, and its white tail is short and wedge-shaped. The species became extinct in Britain in the early twentieth century because of illegal killing, and today's population in the British islands has descended from reintroduced birds. They were first released on Scotland's west coast (and in particular Mull) in the 1970s, after an absence of 70 years. There are now more than 130 breeding pairs in Scotland, but their conservation status remains 'red', meaning they are still in need of urgent protection. (One in four bird species is on the red conservation list in the British Isles.)

The sea eagle was also once widespread along the south coast of England, from Kent to Cornwall, before being driven to extinction. The last breeding pair was seen on Culver Cliff on the Isle of Wight in 1780. Having successfully reintroduced the birds to Scotland, six sea eagles were released on the Isle of Wight in 2019 – part of a five-year programme where at least six birds will be released annually. Each of the birds has an electronic tracker, and within just nine days of being released, a young male called Culver took off on a 680km (422 mile) flight over London and Essex, before returning home to the island. Seven more birds were successfully released in August 2020, and it is hoped they will begin breeding in 2024.

A sea eagle (Haliaeetus albicilla) with a successful catch. When fishing, they fly low over the water, stop and hover, before dropping to take fish from the surface. They are versatile and opportunistic hunters and carrion feeders, sometimes taking food from other birds and even otters. They mainly eat fish, but will also feed on birds, rabbits and hares.

A Guide to Coastal Birdwatching Sites

Grab a pair of binoculars and a good pair of walking boots and take time to get to know the birds around our coastline. It is important to follow a few simple guidelines:

• Obey the law and respect the rights of others.
• Always make the welfare of birds and their environment your priority.
• Do not get too close to the birds.
• Stay on paths where they exist.
• Ensure that nest structures, feeders and artificial bird environments are not disturbed.

If you go in large groups, this has to be organised carefully.

Here are just a few sites where you can find coastal birds in their natural habitat. Many cliff sites are on islands and will involve a boat trip:

Clifftop habitats

• Isle of Muck, County Antrim: just off the Antrim coast in Northern Ireland, this island is a summer breeding ground for fulmars, shags, razorbills, guillemots, kittiwakes and many other species.
• Isle of Eigg, Inner Hebrides: run by the Scottish Wildlife Trust, this is good in late spring for nesting terns, puffins, oystercatchers and snipe. The ferry from Mallaig will stop if whales or dolphins are spotted.
• Lunga, Inner Hebrides: this unpopulated island is described as 'a green jewel in a peacock sea'; it is also home to thousands of puffins. Boats departs from Ulva Ferry, Ardnamurchan, Tobermory and Fionnphort.
• Handa Island, Highlands: this Scottish Wildlife Trust nature reserve reverberates in summer with the sound of puffins, razorbills, guillemots, kittiwakes, fulmars and Arctic terns. Ferry service from Tarbet.
• St Abb's Head, Berwickshire: in late spring and early summer, these rugged cliffs provide a nesting site for guillemots, razorbills and kittiwakes.
• Bass Rock, East Lothian: in late January, the world's largest northern colony of gannets arrive. Later in the year, they are joined by puffins, kittiwakes, guillemots, razorbills, fulmars and shags. Boat trips from North Berwick.
• Farne Islands, Northumberland: from May to July you can see breeding puffins, guillemots, terns, and most likely grey seals as a bonus. Boats trips from Seahouses.
• Flamborough Head, Yorkshire: this promontory is Yorkshire's only chalk sea cliff, and hosts thousands of fulmars, kittiwakes, guillemots and razorbills, as well as migrating species in spring and autumn.

- Skomer Island, Pembrokeshire: this protected National Nature Reserve offers puffins, guillemots, gannets, plus grey seals and rare wild flowers. Boat trips are from Martin's Haven, near Haverfordwest.
- South Stack Cliffs, Anglesey: this reserve is run by the RSPB, and in spring, guillemots, razorbills and puffins breed on the cliffs. The Anglesey Coastal Path passes through the reserve.

Wetland habitats

- Portstewart Strand, County Londonderry: the berries growing wild along the shores of the Bann estuary makes this an important overwintering site for wildfowl in Northern Ireland, as well as a stop-over and refuelling station in autumn for migrating birds.
- Montrose Basin, Angus: this nearly circular, enclosed estuary of the river South Esk is managed by the Scottish Wildlife Trust. The mudflats support large numbers of waders and waterfowl. The area is also designated a Ramsar site, which is a wetland site of international importance.
- Craster to Low Newton, Northumberland: this walking trail gives you a chance to see shoreline birds, such as the oystercatcher and turnstone.
- Blacktoft Sands, Yorkshire: on the south bank of the River Ouse where it widens into the Humber estuary. This is the largest tidal reed bed in England. The reserve has a diverse population of waders, including the black-tailed godwit. Raptors such as the marsh harrier, merlin, peregrine falcon and barn owl can be seen from the accessible trails and hides.
- Wicken Fen, Cambridgeshire: the area offers a variety of habitats, including fenland, marsh and reed beds. Bewick's and whooper swans migrate into the wetlands for the winter. Grebe, cormorants and snipe can be seen, as well as raptors like the sparrowhawk, marsh and hen harriers, and barn owls.
- Blakeney Freshes and Strumpshaw Fen, Norfolk: these are popular places for bird watching, especially in autumn and winter when the fields, salt marshes and mudflats provide respite and feeding grounds for migrating wildfowl.
- Minsmere, Suffolk: this large reserve includes reed beds, lowland heath, acid grassland, wet grassland, woodland and shingle vegetation. More than 300 species of birds have been sighted here. Easterly winds can bring large numbers of migrating birds on passage.
- Sandwich and Pegwell Bay, Kent: a good site for residential and migrating birds which feed on mudflats and sandy areas. Waders include the oystercatcher, spoonbill and dunlin. Tern can be seen nesting during the summer months.
- St Helens Duver, Isle of Wight: geese overwinter between October and March, feeding on eelgrass beds off St Helens Ledges and in Bembridge harbour. The site is also important for migrating wading birds such as dunlin, redshank, sanderling and turnstone.
- Brownsea Island, Dorset: located in the middle of Poole harbour, Brownsea's eastern shore is teeming with activity, especially during the autumn migration. The island attracts large flocks of waders, particularly spoonbills, avocets, sandwich terns and black-tailed godwits. Ferries from Poole quay.
- Morecambe Bay, Lancashire: this is the largest expanse of inter-tidal mudflats and sand in the United Kingdom, and the area is protected by law. The sandflats and salt marshes here are important feeding grounds for a quarter of a million wading birds, ducks and geese.

The little egret (Egretta garzetta) is a small white heron with a wingspan of less than 1m (3.3ft). It first appeared in the United Kingdom in significant numbers in 1989. It is most common in estuaries in Devon and Cornwall, Poole harbour and Chichester harbour, as well as East Anglia. They are gradually increasing their range and are now found as far north as the Wash and north Wales.

PART 3

THE HUMAN COAST

The entrance to Whitby harbour, looking north-west towards Sandsend Ness in the distance. Originally called Streonshalh in the early Medieval period, the Vikings sacked the port and monastery in AD 867, and renamed the town Whitby, from the old Norse for White Settlement. Captain Cook's research vessel, HMS Endeavour, was built here and became the first European ship to reach eastern Australia.

THE COAST AS DEFENCE

The Tower of London, seen from south of the river. This historic castle was founded by William the Conqueror in 1066 to protect London from naval attack from the Thames estuary to the east.

As the sun rose over the English Channel on the morning of 26 August 55 BC, the white cliffs were lit by the pale morning light. Offshore, Julius Caesar scanned the coastline with the eye of an experienced general. He had successfully conquered Gaul (France), and he now led a formidable force of around 12,000 experienced soldiers to invade Britannia. Ahead, local warriors lined the cliff tops, daubed in their fighting colours of blue woad, daring the Romans to land.

Caesar's flotilla had left Portus Itius (near today's Boulogne) the previous day and, after an overnight crossing, the army was ready to come ashore. But the general was cautious about landing on the narrow beaches below the chalk cliffs near present-day Dover, and he ordered his fleet to sail 16km (10 miles) up the coast to Deal, where they anchored off the pebbly beach. Caesar's cavalry was held up in Gaul, so his elite heavy infantry was forced to wade ashore in their heavy armour without any support. The Celts soon caught up with the invaders and attacked with a large force, including warriors on horse-drawn chariots. After fierce fighting, the local chieftains sought a truce and handed over hostages.

Four days later, reinforcements of 500 cavalry soldiers and horses tried to make the same crossing of the Dover Straits but were driven back by bad weather. The gale coincided with an exceptionally high tide, and many of Caesar's ships dragged their anchors and were wrecked on the beach.

The Britons saw their opportunity and mustered their forces. Without cavalry support, the invading force was vulnerable; nor could Caesar send a reconnaissance force inland, since he needed his soldiers to repair the fleet. The Britons assembled their warriors and began new attacks.

With Caesar's army bogged down on the Kent shoreline, word soon got back to Rome. The statesman and philosopher, Cicero, wrote a letter to his brother, Quintus, who was in Deal with Caesar's forces. Cicero's letter clearly illustrated the apprehension the Romans felt about the tides around the British coastline: '*How glad I was to get your letter from Britain! I was afraid of the ocean, afraid of the coast of the island.*'

The Romans had good reason to be fearful for they were more accustomed to the benign weather and insignificant tides in the Mediterranean, where the range is typically only a few centimetres. Compare this with the tidal range in the Dover Straits of 6m (20ft) or more, and you can understand their apprehension.

Once most of his ships were repaired, Caesar ordered his forces to return to Gaul. He tried again the following summer, and once again his anchored ships suffered considerable storm damage. The invaders beached their shattered vessels and worked day and night to repair the fleet. With his ships secured, Caesar marched inland to confront the local tribes, but the Romans were vulnerable and the general decided to withdraw and return to Gaul.

The Iron Age Celts put up a spirited defence of their island, but it was the fierce Channel tides and unpredictable storms which made Caesar realise that his invading army was exposed. It would be nearly 100 years before the Romans tried again.

PRE-HISTORY

All around our shoreline is evidence of more than two thousand years of man-made structures designed to make the most of the defensive potential of the British coastline. Long before the abortive Roman invasion in 55 BC, ancient Britons understood the importance of using coastal features as defence, and the Iron Age people identified landforms that could be developed into effective fortresses. Out came their crude picks and shovels, and they dug ditches and built banks to create hill forts – circular earthworks typically less than 1 hectare (2.5 acres), but occasionally much larger.

Hill forts had many uses: they were used for defence, for social gatherings and as places to trade. Most of these prehistoric sites are found inland, but the Celts also built effective coastal hill forts. Cornwall, in particular, has many hill forts built on promontories. These **cliff castles** were mostly built during the Iron Age, which lasted from about 750 BC to AD 43.

The Rumps at St Minver, for example, was created by digging a double row of banks and ditches at the narrowest part of the headland and raising a third rampart further inland for additional defence. Traces of round houses were found at the Rumps during archaeological excavations in the 1970s; this suggests the site was occupied for nearly three centuries.

Other sites nearby include Lankidden, St Keverne (where more than a hectare of headland is enclosed), and Winecove Point, St Merryn, which consists of multiple banks and ditches running across three spectacular headlands.

In Wales, Anglesey is the place to go to find prehistoric promontory forts, including Caer Idris, Dinas Gynfor, Dinas Porth Ruffydd, Mynydd Llwydiarth, Twyn-y-Parc, and Ynys-y-Fydlyn.

The largest Iron Age stronghold in the British Isles is the promontory fort at the Mull of Galloway in

Treryn Dinas is an Iron Age promontory fort near Treen in Cornwall. The promontory slopes away steeply into the sea on three sides, making it an excellent defensive position. The South West Coast Path runs along the outer rampart.

Wigtownshire, Scotland's most southerly point. Here, earthworks some 400m (1,312ft) long isolates a huge area of around 57 hectares (141 acres) at the eastern end of the promontory. In most places the defensive structure comprises three ditches, with banks up to 4m (13.1ft) wide and 2.2m (7.2ft) high.

The Mull of Galloway, where prehistoric earthworks supplement steep cliffs to create an impressive defensive position. The earthworks here date from the Iron Age, probably around 1,000 BC, and are almost certainly defensive.

The nature of the Kent coastline is very different to that in the West Country, Wales and Scotland which have been discussed. There was no opportunity here for the Britons to build cliff castles as protection against another Roman invasion. Even so, the Romans were aware of the pitfalls of making another attempt, and the geographer Strabo wrote that problems with Caesar's previous invasions were '*on account of the fact that many of his ships had been lost at the time of the full moon, since the ebb tides and the flood tides got their increase at that time.*' Spring tides added another effective coastal defence against invasion.

THE ROMANS

What is not often covered in the history books is that the Romans had plans to invade the British Isles on seven separate occasions. After Caesar's ignominious retreats in 55 and 54 BC, Rome's first emperor, Augustus, planned invasions in 34, 27 and 25 BC, but each time he called them off. In AD 39, the psychotic Emperor Caligula tried again and assembled an impressive force of 200,000 men on the coast of France, only to have them gather seashells instead. Nobody really understands why he aborted the invasion – perhaps he found the task too daunting and, in his deranged state, ordering his men to go beachcombing might have convinced him that he had waged a successful war against the sea.

The Romans eventually mounted a successful invasion under Emperor Claudius, but even then, victory was not guaranteed. In AD 43, Claudius assembled a fleet of 628 ships, more than 20,000 soldiers and 2,000 cavalry horses – a huge invasion force for the period. Again, the fleet got into trouble; first it was becalmed in mid-Channel, then the ships were swept north and dangerously close to the treacherous Goodwin Sands, off the Kent coast. The Romans eventually made a successful landing, this time on the gently sloping sands of Pegwell Bay, just south of Ramsgate. The Celts attacked, but they were taken aback by the sheer size of the Roman army; instead, they wisely retreated to high ground where they could take stock of the invasion fleet which filled the 3km (2 mile) wide bay below them.

The local forces were no match for the Roman legions, and over the course of the following year, the invaders battled inland, storming through inland hill forts. They quickly dominated the tribes of south-eastern England, and by AD 47 they established a new city on the north bank of the River Thames, which they called *Londinium*. However, their main objective was further to the west, for the main attraction in Britain (apart from slaves and good hunting dogs), were precious metals found in Cornwall and Wales.

The Romans built new forts, harbours, settlements and roads. However, they had little need to build coastal defences, simply because there was no real threat of an invasion. At least this was the situation until the empire began to weaken, and from the third century AD onwards, the Romans began to construct coastal forts to defend Britannia from repeated raids by Saxon and Frank pirates. These forts became the new frontier to protect the southern and eastern coastlines of the British islands and, if you know where to look, there are still traces to be found around this part of our coastline.

Dubris, also known as *Portus Dubris* or *Dubrae*, was a Roman port built on the site of present-day Dover in Kent, and an important naval base on the south coast for much of the second century AD. Here, the Romans constructed a fort, harbour installations and two lighthouses – one of which remains in the grounds of Dover Castle. By the end of the third century AD, the Romans had built a shore fort on the west bank of the Dour estuary. Excavations have revealed walls 2.4m (8ft) thick with several external towers. The Dour estuary at Dover was a natural harbour for the Roman town, but like many places along the south coast, the river silted up during the Medieval period.

The Romans also built the port of *Rutupiae* or *Portus Ritupis* (today called Richborough Castle), close to Sandwich in Kent. Other coastal forts include *Branodunum* near Brancaster in Norfork (built around AD 230), and *Gariannonum* (Burgh Castle) built around the late third and fourth century near Great Yarmouth in Norfolk. *Othona* was a Roman shore fort near the current village of Bradwell-on-Sea in Essex, built in the third century. All these sites still display evidence of their structure and walls.

The Roman fort at *Anderitum* in East Sussex was later converted into a Medieval fortress, Pevensey Castle. The fort used a peninsula rising above the coastal marshes as part of its defence. The early structure was built around AD 290, based on the dating of wooden piles. After the collapse of the Roman Empire, there is evidence that the local population moved into the abandoned fort, perhaps for protection against Saxon raiders. Today, the castle is left stranded 1.3km (0.8 miles) inland – again a testament to the problem of silting along the south coast of England since the Roman period.

The Roman lighthouse and the heavily restored Saxon church of St Mary in Castro both lie within the confines of Dover Castle.

William the Conqueror's castle at Pevensey was built on the site of one of the last and strongest Roman 'Saxon Shore' forts. The castle was undefended in 1066, and it was here that William's army came ashore; the harbour also provided a safe anchorage for his fleet. The bay has since silted up, and the castle now stands 1.3km (0.8 miles) from the coastline.

The best-preserved Roman fort north of the Alps is *Portus Adurni*, situated at the inshore end of Portsmouth harbour. The fort was built during the third century and encloses an area of 3.6 hectares (8.9 acres), with outer walls 6m (20ft) high. After the collapse of the Roman Empire, *Portus Adurni* became an Anglo-Saxon residence, and later a Norman castle and Medieval palace. The site was occupied for almost 16 centuries, with its last official military function being a jail for captured French soldiers during the Napoleonic Wars.

Portus Adurni overlooks Portsmouth harbour, and was later converted into the Medieval Portchester Castle. The twelfth century church was built within the outer bailey and is now the parish church.

THE MEDIEVAL PERIOD

After the Romans withdrew from Britannia in AD 410, the English coastline was left unprotected. From about AD 430 to 500, Saxons from Germany, Jutes from northern Denmark and Angles from southern Denmark crossed the North Sea generally unopposed and settled in the fertile lands of 'Angle-land' or England. Over the next 100 years or so the invading kings established their realms, which survive to this day as English counties and regions: Kent (Jutes), Sussex (south Saxons), Wessex (west Saxons), Middlesex (middle Saxons), East Anglia (east Angles) and Northumbria (land north of the Humber).

The Anglo-Saxons are known to have cremated their dead and used their boats in funeral rituals. The best burial mound of the period is undoubtedly at Sutton Hoo, on the Deben estuary near Woodbridge in Suffolk. The grave contained the remains of a vessel 27m (89ft) long and was excavated in the late 1930s; it is believed to be that of a king of East Anglia, dating from the early seventh century. The site enhanced our understanding of the early Medieval period or 'Dark Ages', and the elaborate gold and silver artefacts found in the grave suggests there were trading links with Scandinavia, north Africa and the eastern Mediterranean. These relics can now be seen in the British Museum.

Later, in the eighth century, a lack of any significant coastal defences contributed to the ease with which the Vikings attacked and penetrated Anglo-Saxon kingdoms. Their first raid was in AD 793 at Lindisfarne in Northumberland and continued through to the late eleventh century. Initially, the Viking attacks were localised 'smash-and-grab' raids, but they soon adapted their tactics and took their swift longboats up estuaries and rivers, including the Tyne, Humber, Ouse, Thames and Severn, where they were able to penetrate far inland, much to the indignation of the local inhabitants.

The Viking Age ended in AD 1066, which was the last time the British Isles were invaded by a foreign force. The Normans landed at Pevensey in Sussex, and William of Normandy immediately ordered three fortifications to be built at Pevensey, Hastings and Dover. Today, the walls of William's coastal castles are

in varying states of disrepair, but all three still stand as a testament to the importance of the coastline in defending the nation.

The Holy Island of Lindisfarne in Northumberland; a monastery was first founded here around AD 634, but the community was sacked during a raid by Norsemen in AD 793. This attack is generally taken to be the beginning of the Viking Age in the British Islands.

There was probably an Iron Age hill fort at Dover before AD 43. The Romans then built their own garrison at Dubris around AD 113, and it was subsequently rebuilt by William in AD 1066. Dover Castle was still in use during the Second World War, where a network of top-secret tunnels became the headquarters for the rescue of troops from Dunkirk. The structure is still intact and open to the public.

From that time onwards, English and then British monarchs were acutely aware of the importance of national defence. About 1,000 years ago the great age of castle building began, which lasted for nearly 500 years. It is estimated that about 4,000 castles were built in England alone during the Medieval period, with around 300 still surviving in some recognisable form, many along the British coastline.

Like the Iron Age Celts and the Romans before them, the Medieval castle-builders used pre-existing landforms to good effect. Tintagel Castle in Cornwall is, perhaps, one of the most famous coastal castles in Britain – famed as the legendary birthplace of King Arthur. The site was almost certainly occupied during the Roman period, but the Medieval stone structure dates from the thirteenth century.

Hastings Castle was originally a wooden tower built on top of a mound or motte, surrounded by an outer courtyard or bailey, which was enclosed by a wooden palisade. After William's coronation on Christmas Day 1066, he issued orders for the castle was to be rebuilt in stone.

Tintagel Castle in Cornwall used a steep promontory to assist defence.

The Normans went on a building spree, and it is believed that at least 500 – and possibly as many as 1,000 castles – were built by the end of the eleventh century; this was barely two generations after the Norman landing. The best-preserved Norman castle is undoubtedly the Tower of London which William began in late 1066. Like the Romans, he well understood the strategic advantage of defending the estuarine entrance to London, and the castle remained an effective bastion for hundreds of years and was still used as a prison and site of a dozen executions during the First and Second World Wars.

Not all of William's castles were built on the coast of course, but he knew that the British islands remained vulnerable to foreign invasion for as long as the coastline was left unguarded. He was determined to learn the lesson of history and not allow this to happen again.

What all these castles have in common is the use of a promontory or marshland to give additional protection from attack, while defending a strategically important stretch of water.

Carrickfergus Castle is surrounded on three sides by the sea.

Dunnottar Castle is built on a rocky headland, about 3km (2 miles) south of Stonehaven, on the north-eastern coast of Scotland. The site dates from the early Medieval period, although the surviving buildings were mainly built in the fifteenth and sixteen centuries. The Scottish crown jewels were hidden here from Oliver Cromwell's invading army in the seventeenth century. The castle is now restored and open to the public.

Medieval castles were generally fortified residences, but as the threat of invasion increased, so too did the coastal defences. The forts around our coastline tell the stories of the country's defiance against a range of enemies over more than five centuries: French, Spanish, Dutch and German. Over the years, many defences were updated to face successive foes wielding increasingly sophisticated weapons, ranging from early Medieval cannons to Second World War anti-aircraft guns.

Dartmouth Castle, for example, was built in response to the threat from the French, and the building dates from the 1380s. Just 400m (1,312ft) away on the opposite side of the Dart estuary is Kingswear Castle, built between 1491 and 1502. Together, these fortifications defended the entrance to the Dart estuary and the strategic town of Dartmouth – an important deep-water trading port and a home to the Royal Navy since the reign of Edward III. Even with primitive guns firing stone cannon balls (which had a range of only a couple of hundred metres), these two artillery castles ensured the residents of Dartmouth could sleep easily at night.

Dartmouth Castle and Kingswear Castle (foreground) in the late nineteenth century. The forts defend the entrance to the Dart estuary. These castles acted as a double defensive system as a chain was extended across the river between the two forts to create a barrier.

It was around the thirteenth century that England began to develop a proper navy to supplement the land-based fortifications. The navy had no defining moment of formation but started as a motley assortment of 'King's ships' during the Middle Ages, brought together when needed and then dispersed when the threat had passed. King John ordered the construction of 54 royal galleys between 1207 and 1211 and, as an embryonic Royal Navy developed, a naval base was established in Portsmouth.

THE MODERN ERA

It was Henry VIII who really understood the importance of maritime defence, and he created a standing 'Navy Royal', with dockyards and a permanent fleet of purpose-built warships. He also built a chain of 30 coastal fortifications between 1539 and 1547 called 'Device Forts'. These Tudor forts protected England's south and east coasts against the threat of invasion from Catholic Europe. They were characteristically squat and rounded to deflect incoming cannon balls and provide all-round firepower for their own heavy guns. A good example is Deal Castle in Kent.

Deal Castle was built by Henry VIII between 1539 and 1540 to defend the strategically important Downs anchorage off the Kent coast. The moated castle had 66 firing positions for artillery. Henry improved Dover Castle at the same time, for the same reasons.

Henry VIII's great flagship, the *Henry Grace à Dieu* ('Henry, Thanks be to God') dates from around this period. It was an English carrack or 'great ship', typically very top heavy, with a large forecastle four decks high, 50m (164ft) long, and a crew of 700 to 1,000 men. Henry's flagship was present at the Battle of the Solent against French forces in July 1545, when the *Mary Rose* sank so spectacularly barely a mile off the entrance to Portsmouth harbour.

The wreck of the *Mary Rose* was discovered in May 1971 and raised in 1982. Much of the ship was preserved in the fine mud, together with many items on board, including carpenters' tools, medicine flasks and a large number of wooden dishes. Also found were nit combs (complete with nits), and 655 items of clothing. The wreck of the *Mary Rose*, together with the artefacts found on board, are exhibited in the Portsmouth Historic Dockyard.

The threat from continental Europe continued after Henry's reign, but the next attempt at a full-scale invasion came not from the French, as feared, but from the Spanish. In the summer of 1588, 130 of Spain's finest fighting ships sailed up the English Channel, intent on an invasion and deposing the Protestant Elizabeth I of England. The Spanish armada was first sighted off the Lizard Point in Cornwall, and a chain of fire baskets was lit. Each beacon was built within sight of another, allowing a warning of the approaching threat to reach London within just 30 minutes. Only one stone signal beacon survives, and it is found not on the coast but 80km (50 miles) inland, just north of the village of Culmstock in Devon.

The Henry Grace à Dieu was launched in 1514. The ship carried 43 heavy guns and 141 light guns and was the largest and most powerful warship in Europe.

A stone a signal station at Culmstock was built in 1588 to support fire baskets which were lit to warn of imminent invasion. This was one of a chain of signal stations stretching from Land's End to the rest of the country.

The Spanish fleet was famously overwhelmed when anchored off the French coast, and the survivors sailed north into the North Sea. Of the original 130 ships in the fleet, only 67 returned safely to Spain. Almost as many ships were lost to bad weather or on the rocky shore of 'fortress Britain' as were sunk by the English navy. Once again, the weather, tides and coastline of the British islands proved to be a formidable defence.

From the mid-eighteenth century, Britain supported the world's most powerful navy. However this did not mean that coastal defences were redundant, for the threat from the French was still very real. Between 1804 and 1812, the British government built 103 Martello towers around the coastline of southern England, stretching from Seaford in Sussex to Aldeburgh in Suffolk. These fortified structures typically had two floors, stood about 12m (40ft) high, and housed an officer and 15 to 25 men. Their round shape and thick walls made them resistant to cannon fire from ships, and their height made them an ideal platform for a single heavy cannon mounted on the flat roof, able to rotate and fire through 360 degrees. The towers became obsolete with the introduction of more powerful artillery, but many survive either as private homes or preserved as historic monuments.

By the middle of the nineteenth century, the risk of attack by the French was still very real, and in 1860 a Royal Commission recommended that the defence of the south coast should be improved – although there was heated debate in Parliament whether the cost could really be justified. In the end, the government got its way under the premiership of Lord Palmerston, who was a strong supporter of the proposals.

The result was a series of coastal forts which became popularly known as 'Palmerston's Follies' – suggesting a costly ornamental building with little practical value. This moniker was not altogether surprising, as the first ones built in the Solent bizarrely had their main armament facing inland to protect Portsmouth harbour from a land-based attack. The criticism was also justified because, by the time the forts were completed, the threat from the French had passed, and advances in artillery made them ineffective. In the end, these Victorian coastal forts became the most costly and extensive system of fixed defences ever undertaken in Britain during peacetime.

Spitbank Fort in the Solent was built between 1861 and 1878. It fell into disuse between 1956 and 1982 and was declared a Scheduled Monument in 1967. It has now been converted into a boutique hotel and restaurant.

This Martello tower in Shingle Street, Suffolk was built in the early 1880s. During the Second World War they were used as observation platforms or housed anti-aircraft guns. Today many are refurbished for residential use or holiday lets.

1914 brought the outbreak of the First World War and the fear of a German invasion. Many Victorian coastal forts were modernised, and the first concrete pillboxes in England were constructed at likely invasion points in Norfolk, Essex and Kent. A 9-inch battery was built at Spurn Point in East Yorkshire, but the site has since been lost to coastal erosion. Coastal artillery batteries were also built to defend the Tyne, Humber and Orwell estuaries, as well as the submarine base at Blyth, Northumberland. The nearby Blyth Battery still survives virtually intact, as do two remarkable gun towers at Sheerness in Kent.

Towards the end of the First World War, the British Admiralty became worried about the threat of German submarines passing from the North Sea and into the English Channel. Eight towers were designed – code named M-N – to be built across the Straits of Dover to protect allied merchant shipping. By the end of the war, only one tower had been completed at the huge cost of £1 million and was left sitting in Shoreham harbour awaiting deployment.

In 1920, the completed tower was towed out to the Nab rock off the Isle of Wight, at the entrance to the eastern Solent. The tower was flooded and settled on the seabed at an angle of 3 degrees from the vertical – a characteristic tilt which can still be seen today. The Nab Tower was manned as a lighthouse, and also functioned as a Royal Navy signal station. During the Second World War, it was equipped with a pair of Bofors guns to provide defence in the Solent, and the crew succeeded in shooting down several enemy aircraft. The Nab Tower still functions as a lighthouse but has been unmanned since 1983.

The outbreak of the Second World War saw the widespread construction of a large number of substantial coastal defences. After the Dunkirk evacuation in May and June 1940, the question Britain asked itself was not 'will Hitler invade?', but rather 'when?' The German invasion plan was called Operation Sea Lion, and the threat provoked frantic construction of defences throughout Britain. The most vulnerable areas were the east and south coasts of England, and here emergency coastal batteries were built to protect ports and possible landing places.

The Nab Tower being refurbished in 2013, when the height of the structure was reduced from 27m to 17m (89ft to 56ft). (Taken from Solent Cruising Companion)

By the end of 1940, thousands of pillboxes, anti-tank structures, coastal defences, heavy-gun emplacements and anti-aircraft batteries had sprung up all over the British Isles. Most of these fortifications were fitted with whatever artillery pieces that could be found – mainly from naval vessels which had been scrapped after the First World War.

The most common fortifications were Type 22 and 24 hexagonal pillboxes, and the square Type 26. These were part of the most extensive system of coastal home defence since Napoleonic times, and many still remain – with the majority found along Britain's coastline. They can still be seen today, crumbling and overgrown, and a poignant reminder of Britain during its darkest hour.

Pyramid-shaped concrete blocks can still be found around the coast, designed to force advancing tanks to lift, thereby exposing vulnerable parts to gunfire. Nicknamed 'dragon's teeth', they varied in size but were typically 60cm (2ft) high and about 90cm (3ft) at the base. Other anti-tank obstacles which have stood the test of time are 'hairpins' – lengths of bent railway line concreted into the ground and designed to perforate the underside of a tank.

A hexagonal pillbox on Slapton Sands in Devon. The foundations were built on bedrock, and this has allowed the structure to withstand decades of weathering intact.

Studland Bay in Dorset was vulnerable to a mechanised invasion force. Here, anti-tank blocks of 'dragon's teeth' are set just behind the sandy beach as an effective defence.

The coastal artillery battery on Culver Down on the Isle of Wight overlooks the eastern Solent and Nab Tower. The fortification was one of the original Palmerston Forts built on the island and served in two World Wars until it was closed in 1956.

Pillboxes and coastal batteries were active fortifications designed to repel potential invaders. However, there were also more passive defences, designed either to detect the enemy or to make life difficult should an invasion be attempted. Access to many beaches along the coastline of England was blocked with barbed wire, which frequently marked extensive minefields. Fortunately, these have long been removed.

Romney Marsh was the planned invasion site of Operation Sea Lion, and the wetlands here were flooded to prevent invasion, as well as low-lying areas on the coast in East Anglia. The south coast of England had large numbers of seaside piers, which were ideal for landing troops. These were disassembled, blocked or otherwise destroyed, and many were not repaired until the late 1940s or early 1950s (see chapter 13).

Some of the most unusual Second World War coastal features were acoustic mirrors, which are passive concrete structures designed to reflect and focus sound waves. They were crude but effective parabolic microphones, and during the Second World War a network of these devices was ordered to be built. However, the project was stopped when electronic radar was developed; even so, several are still standing today.

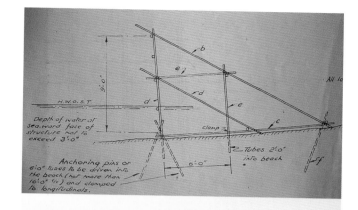

An acoustic mirror overlooking the English Channel at Abbot's Cliff, between Dover and Folkestone.

The giant 60m (200ft) acoustic mirror at Denge in Kent.

The remains of Admiralty scaffolding (see design above) can be found all around the south and east coast – this was at Bawdsey Quay in Suffolk. They were simple coastal deterrents – steel tubes were sunk into beach sediment to slow potential invasion forces.

The demands for national defence have changed in the age of sophisticated radar, cruise missiles, stealth bombers and tactical nuclear warheads. There are still naval dockyards at Rosyth in the Firth of Forth, in Faslane on the River Clyde in Scotland and in Plymouth and Portsmouth on the south coast of England. But today, the coastal defence of the British Isles is no longer designed to deter an invasion, but mainly to prevent the illegal smuggling of drugs, arms and people into the United Kingdom.

UK Border Force HMC Eagle coastal patrol vessel entering Plymouth harbour. The Border Force is a law-enforcement command within the British Home Office and operates independently of the Royal Navy. The force is responsible for frontline border control at air, sea and rail ports in the United Kingdom.

Six hundred years of naval defence: as HMS Queen Elizabeth leaves Portsmouth, the ship sails past the Round Tower built by King Henry V in 1481 to defend the harbour. The aircraft carrier is the biggest warship ever built for the Royal Navy and represents four acres of sovereign territory ready for deployment anywhere in the world.

CHAPTER 12

THE COMMERCIAL COAST

As an island nation, overseas trade has been important in the Britain Isles for more than 2,000 years.

In the summer of 2002, archaeologists working in Poole harbour made a remarkable discovery. They found thick tree trunks with pointed ends driven deep into the mud as foundations for two jetties; one was 55m (180ft) long, and the other a remarkable 160m (525ft) long; both jetties were originally covered in slabs of Purbeck limestone. The evidence suggested that the jetties extended out from the shoreline to a deep-water channel and allowed boats to sail into the sheltered water of Poole harbour and tie-up alongside at a well-built quay.

Perhaps the most extraordinary thing about these structures was the date: they were built around 250 BC, more than two centuries before the first Roman invasion. Artefacts found in the mud came from a nearby Iron Age settlement at Cleavel Point on the shoreline of Poole harbour, and it showed that maritime traders sailed from France to barter for pottery, shale jewellery and other things made locally in Dorset.

The archaeologists realised they had unearthed Britain's first major cross-channel port, and it was more than two thousand years old.

COASTAL TRADE

Our little corner of north-western Europe is one of the world's top ten archipelagos, with more than 6,000 islands stretching for 1,300km (808 miles) from the Isles of Scilly in the south-west to the Shetland Islands in the north. Just as the ancient Greeks and Romans were able to extend their horizons by sailing around the eastern Mediterranean, so the early Britons took to their own waters (albeit in more modest vessels).

The weather and tides were just as testing 2,250 years ago as they are today, so the Celtic mariners relied on the protection of natural harbours such as Poole. Hengistbury Head in Dorset is another good example, and the area is known to have been inhabited 12,000 years ago when early hunter-gatherers moved into southern England as the great ice sheet began to retreat. There was a prospering community here during the early Bronze Age, and axes and funeral urns have been found to be up to 4,000 years old.

By the late Iron Age (around 2,000 years ago), cross-channel trade from both Hengistbury and Poole was flourishing, with iron, silver and bronze traded for figs, tools and other goods. Because of its strategic importance, the promontory at Hengistbury was fortified with earthworks to create a cliff castle and, when the Romans arrived, they were quick to recognise the advantage of the harbour. Soon exotic goods such as wine and glassware were being shipped from Gaul (France), and metals, slaves, corn, cattle, hides and dogs were traded in exchange. Earthworks and barrows can still be traced around the headland but, if you go exploring, do beware of adders in the sandy areas.

The south coast of England, however, is in a constant state of change. Over the centuries, longshore drift has brought sand and shingle up the coast and created a longshore spit, which has partially blocked the entrance to Christchurch harbour, making it impractical to use as a trading port.

The view from Hengistbury Head, looking east towards the village of Mudeford. On the left is Christchurch harbour, in the middle is the sandy spit formed by longshore drift.

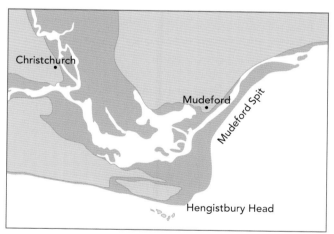

Christchurch harbour on a map dated 1867. Longshore currents have deposited sediment along the shoreline in a north-easterly direction and blocked the harbour entrance to all but the smallest boats.

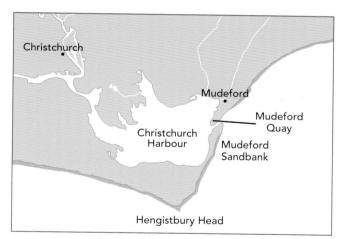

Christchurch harbour today, showing the changing entrance.

Similar problems occurred further up the coast in East Sussex and Kent, where the coastline has changed dramatically over the centuries. When the Romans landed in Pegwell Bay in AD 43, the Isle of Thanet really was an island, isolated from the rest of Kent by the Wantsum channel. The channel gradually silted up, and the last ship sailed through the passage in 1672.

In the eleventh century, Edward the Confessor established the confederation of Cinque Ports of Sandwich, Dover, Hythe, New Romney and Hastings to furnish his private navy. Over the years, these five strategic ports have all silted up, and the huge cross-channel ferry port at Dover only exists because of massive harbour walls.

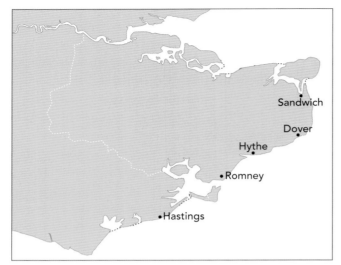

The Cinque Ports in 1287, which supplied ships and men for the king's navy until the fourteenth century. Today, silting has completely closed all these harbours.

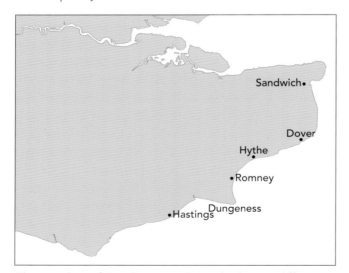

The coastline of East Sussex and Kent looks very different today.

Despite the setbacks of silting in southern England, the British coastline and estuaries became essential for trade, just as railways, motorways and airports are today. By the late Medieval period – roughly between 1250 and 1500 – the British islands were littered with small harbours all around the coastline, and up into estuaries and creeks. Trading ships were small and relied solely on the wind and tides; voyages could take weeks, but it was still quicker, easier and safer than using the roads, which were little more than dirt tracks.

A Medieval ship dated to between 1447 and 1449; the masts are not shown. These trading ships, called cogs, were typically 15m to 20m (49 to 66ft) long, and traded between continental ports. Local coastal vessels were usually smaller.

Navigation was an imprecise art and cargo ships hugged the coastline when they could, but longer open water crossings were made when necessary, and maritime trade from and around the British islands was extensive. Cornish tin was shipped to the continent, and luxury items like dates came from as far away as north Africa. From the fifteenth century, wool and cloth produced in England increasingly dominated continental markets, and harbours along the east coast such as Kings Lynn prospered from being close to the centres of English weaving.

Charlestown, near St Austell in Cornwall is a pristine example of a late Georgian port. Built between 1791 and 1801, it served the burgeoning mining industry and was originally used to export copper and import coal, then later china clay. Today it is home to working ships, and frequently appears in television and film period dramas.

Boscastle is typical of the small harbours which grew up all around the British Isles. The stone quay was built during the Elizabethan period and has hardly changed since. The harbour was once one of the busiest in north Cornwall, and it was used to export slate and local produce, and to import limestone and coal. It is still a working fishing harbour.

The spritsail Thames sailing barge May was built of wood in Harwich in 1891. The flat-bottomed boat is 24.7m (81ft) long, 5.8m (19ft) wide, but has only a shallow draught of 1.2m (4ft).

The demands of coastal trade sometimes led to innovative solutions. The coastline of East Anglia and north Kent is broken by a variety of estuaries, which offer good access inland. As a result, the Thames sailing barge evolved, and were once a familiar sight all along the east coast and Thames estuary. Their flat bottoms and shallow draft allowed them to sail far up an estuary and dry out on the mud to unload at low tide. They were used to transport anything from cement and bricks to grain and sugar beet, and some of the bigger vessels brought coal from Newcastle to shallow ports along the Thames estuary. The unique spritsail sailing rig traditionally allowed the vessel to be sailed by a man and a boy; originally none of the boats had engines. There were more than 2,000 barges registered in the early twentieth century, but by 1954 only 160 barges were still trading. Improvements in road transport and the development of container shipping changed the way goods were transported.

'Mist In Port' by Charles de Lacy (1856-1929) and dated 1881. The painting of the Pool of London shows the transition from sailing ships to steamships towards the end of the nineteenth century.

All around the coast are buildings which tell much about the history of the area. The first lime kilns at Beadnell on the Northumberland coastline were built in 1789 and are now Grade II listed. Limestone was burnt in the kilns to make quicklime, used as a fertiliser and for brick mortar. By 1822, the kilns had fallen out of use, and they are used today by local fishermen to store lobster pots.

The introduction of containerisation in the second half of the twentieth century revolutionised international shipping and brought to an end the extensive use of traditional general-purpose cargo vessels. As international trade grew through the 1980s and imports increased (especially from China), so container ships became larger, and they demanded bigger, deep-water ports. The HMM *Algeciras* is currently the largest container ship in the world (as of 2021), with the capacity to carry the equivalent of 24,000 6m (20ft) containers.

London was once the largest port in the world, but it was far up a long estuary – not ideal for the new container ships. Inevitably, the wharves began to disappear: some were demolished, but some docks, such as St Katherine docks near Tower Bridge, have survived as moorings for private boats. Downriver on both sides of the Thames, old warehouses have been converted into luxury riverside flats. This re-purposing of Victorian working buildings can be seen throughout the country, from Glasgow to Newcastle and from Liverpool to Bristol.

Felixstowe Docks (right): Felixstowe in Suffolk is now the busiest container port in the UK, handling nearly half of Britain's containerised trade. The port is capable of handling 3,000 ships a year. The shipping can be observed from Harwich or Shotley on the opposite bank.

Like many dockland warehouses around the country, industrial buildings in London are being converted into luxury flats.

ROADS & RAILWAYS

The rise of Britain as a global trading nation coincided with the Industrial Revolution, which began around 1760 and ran through the first half of the nineteenth century. First canals, and later railways, connected cities to manufacturing regions and the coast. In 1830, the Liverpool and Manchester Railway opened, leading the way to a boom in railways carrying both freight and passengers. A new transport age had begun.

The British coastline is incised by numerous large estuaries – the Seven, Mersey, Clyde, Tyne, Humber and the Thames, to name just a few. From the 1830s, railway engineers grappled with the challenge of how the railways could cross these wide river mouths. Over the next couple of decades, railway tracks, embankments, viaducts and bridges began to change the face of the British coastline.

The oldest railway bridge in the world that is still in use was built in 1825 to cross the River Skerne in Darlington. After the Liverpool and Manchester Railway opened in 1830, the expansion of the railways was dramatic – and with it came the challenge of crossing the estuaries. The wooden Barmouth bridge was opened in 1867, and carries the Cambrian coast railway across the wide, shallow Mawddach estuary. With the exception of the section which swings open to allow tall ships to pass, the rest of the bridge is timber trestles. It is now a Grade II* listed structure.

The Barmouth bridge over the Mawddach estuary is built mainly from timber trestles. The bridge carries rail, cyclists and pedestrians, and is part of the National Cycle Network. Most of the bridge is built on a gravel bed which is covered with shifting sand. Water flows through the deep-water channel at up to 9 knots (16.7km/h).

The Britannia bridge across the Menai Strait was built by Robert Stephenson in 1849; the box girders for the main 140m (460ft) span weigh 1,900 tonnes. The bridge connects Anglesey to the mainland and was an important part of the rail link between London and Holyhead, with its ferry service on to Dublin in Ireland. In 1970, a fire caused extensive damage and it was necessary to completely rebuild the structure in a different configuration. The rebuilt bridge now supports two decks, carrying both road and rail.

The railway bridge over the River Forth in Scotland was completed in 1890. It was voted Scotland's greatest man-made wonder in 2016 and is now a UNESCO World Heritage Site.

Britain's Most Spectacular Coastal Railways

Although many local lines were closed in the 1960s, there is still an abundance of railway routes through spectacular scenery, which give us a unique perspective of our coastline. For steam enthusiasts, special excursions are periodically on offer. These are some of the best:

1. Exeter to Teignmouth, 25 minutes: the train does not leave the coast between Teignmouth and the Exe estuary, giving passengers fabulous views of the south Devon coastline.

2. St Ives to St Erth, 15 minutes: the tracks follow Carbis Bay before passing the Hayle estuary.

3. Norwich to Sheringham, 1 hour: combine East Anglia's rural landscape with stops at two of the region's best holiday resorts – Cromer and Sheringham.

4. Swansea to Pembroke Dock, 2 hours: the line snakes through south Wales and stops at stations giving access to the Pembrokeshire Coast Path.

5. Rhyl to Holyhead, 2 hours: through Llandudno, Bangor and over the Menai Straits to the island of Anglesey.

6. Machynlleth to Pwllheli, 2.5 hours: the Cambrian Line runs tight around Cardigan Bay with potential stop-offs at Aberdovey, Fairbourne and Barmouth. Connect at Porthmadog for the narrow-gauge train to Blaenau Ffestiniog.

7. Barrow-in-Furness to Carlisle, 3 hours: The track follows the sea for most of the journey and you can stop off at the seaside towns of Ravenglass, Silecroft and St Bees.

8. Middlesbrough to Newcastle, 1+ hours: parts of the Durham Coast Line first carried passengers in 1835, and the trains run close to the coastline giving wonderful views.

9. Durham to Berwick-upon-Tweed, 1 hour: the railway runs along the cliffs of the stunning Northumberland coastline.

10. Edinburgh to Dundee, 1.25 hours: a wonderful chance to pass over the iconic Forth bridge and get a spectacular view down onto the Firth of Forth.

11. Inverness to Thurso, 4 hours: the most northern railway line in the UK runs along the Firths of Beauly, Cromarty and Dornoch, before passing along the North Sea coast.

12. Derry to Portrush, 1 hour: lots of tunnels and spectacular scenery, plus the chance to take a steam train in summer to the Giant's Causeway.

The railway line outside Teignmouth, looking north across Tor Bay to Exmouth. This section is part of one of the great coastal railway lines in the country.

From the 1960s, road traffic in the United Kingdom increased dramatically from the growth in car ownership, an expansion in international trade, and the use of more and more shipping containers. This inevitably led to a dramatic increase in new road bridges across estuaries throughout the country.

In Plymouth, Isambard Kingdom Brunel's Royal Albert Bridge was completed in 1859 to carry the new railway into Cornwall. It spanned the River Tamar alone for more than a century before the Tamar Bridge was built in 1961 to give an alternative road crossing to the quaint but slow Saltash and Torpoint ferries (still in use today and used primarily by local traffic).

The increase in motorways required new bridges to carry multiple traffic lanes. A fixed link across the Severn estuary was first proposed as far back as 1824 to improve the mail coach service between London and Wales, but it was not until 1966 that a road bridge finally connected the principality, replacing the Aust ferry. In 1996 – just 30 years after the construction of the first bridge – a second Severn crossing was opened due to the increase in traffic between England and Wales. The second Severn crossing marks the lower limit of the River Severn and the start of the Severn estuary.

The original Severn Bridge was opened in 1996 and was 1.6km (0.99 miles) long. The suspension bridge was granted Grade I listed status in 1999, just 33 years after it was finished. As road traffic increased, a second Severn crossing was built. The Prince of Wales bridge was opened in 1996, but a more complicated route caused the bridge to be 5.1km (3.2 miles) long.

The Royal Albert Bridge carries the railway line over the River Tamar between Plymouth in Devon and Saltash in Cornwall. It was designed by Isambard Kingdom Brunel and opened on 2 May 1859. Beyond is the new Tamar Bridge, which was opened to the public in October 1961.

The Humber Bridge was built with a single span because of strong tides and shifting sands. It opened in 1981, and for 17 years it was the longest single-span suspension bridge in the world.

The Queen Elizabeth II bridge across the Thames estuary, from the south bank of the river at Greenhithe in Kent. The crossing comprises of two bored tunnels and the cable-stayed bridge. This is the only fixed road crossing of the Thames east of the city, and it has become the busiest estuarine crossing in the country, averaging more than 130,000 vehicles a day.

FISHING

The United Kingdom has a long and proud history of sea fishing, both in home waters and further offshore. Excavations at the Iron Age hill fort at Broxmouth in East Lothian, Scotland during the 1970s found fish bones from large specimens of ling and other deep-water species. Ling are most commonly found at depths between 100 and 400m (330 and 1,310ft), although young fish are found in shallower water. This unexpected evidence suggests that these prehistoric people were engaged in deep-sea fishing more than 2,000 years ago using simple dugout and skin boats.

Until Henry VIII's English Reformation in the 1530s, Catholic Britain had a voracious appetite for fish on Fridays. Some came from inland rivers and lakes, but other sources included marine fish from inshore waters and further offshore. British boats are known to have fished Icelandic waters since the fourteenth century. There is also plenty of evidence that fishing boats crossed the Atlantic in the early fifteenth century in search of abundant cod in Newfoundland waters – decades before Columbus sailed for the New World. It was around this time that the Dutch developed a type of offshore vessel called a herring buss, which became a blueprint for European fishing boats. These bluff-bowed sailing ships were typically 15m (50ft) long, and crewed by about two dozen men, who sometimes stayed at sea for weeks on end.

The oldest known drawing of a herring buss, dated to around AD 1480.

As the country's population and wealth grew during the nineteenth century, so too did the demand for seafood. Without refrigeration, fish traditionally had to be salted or dried. But the growth in the railways allowed fresh fish to be caught and shipped quickly inland to town and city centres. Steam trawlers began to replace sailing boats from the 1870s onwards. These vessels had a larger hold capacity, trawled at greater depths, and used bigger nets.

In 1925, Clarence Birdseye invented a system of freezing fish at sea, and this allowed long-distance fleets to supply freshly frozen fish to British consumers. This breakthrough, combined with the introduction of factory trawlers in 1950, resulted in a significant increase in catches. It was the beginning of fishing on an industrial scale, and it inevitably led to competition for fish stocks.

Rivalry between fishing nations is not new. In the sixteenth century, Dutch fleets of 400 to 500 ships were fishing over the Dogger Bank and around the Shetland Isles. These boats were usually escorted by Dutch naval vessels, because the English considered the vessels to be 'poaching' in waters they claimed were their own. British fishing boats were just as guilty, and intermittent disputes over fishing rights between Britain and Iceland in the north-east Atlantic go back as far as the fifteenth century. The only way these long-running disputes could be resolved was at an international level.

It took until 1994 before a nation's territorial limits were finally agreed after interminable squabbling. The United Nations agreement established that a coastal state is free to set laws, regulate use and exploit any resource within a boundary of 12 nautical miles (22km or 13 statute miles) from their shoreline. Foreign ships were granted the right of 'innocent passage' through these waters; however, fishing, polluting, weapons practice and spying are not considered to be 'innocent', and submarines are required to navigate on the surface and fly their maritime flag.

An Exclusive Economic Zone (EEZ) was also defined as the area offshore in which a sovereign nation has special rights to explore and exploit marine resources, including energy production from water and wind.

The EEZ stretches from the shoreline to 200 nautical miles (370km) offshore. The main difference between territorial waters and the EEZ, is that the first confers full sovereignty over the waters, whereas the EEZ is a 'sovereign right' to exploit resources below the surface of the ocean – the surface waters in the EEZ remain international.

The establishment of these exclusive zones has been very important for the United Kingdom. Between 1973 and 2020, the country was a member of the European Economic Community (EEC), later evolving into the European Union (EU). Whilst a member of the EU, marine resources were shared between member states to some extent. Since the UK left the EU, the country is able to reclaim sovereignty over its coastal waters. The separation agreement with the EU gives member nations residual access to fishing rights in UK waters, but these will gradually change and evolve over time. However, the area of coastal waters over which UK now has sovereignty is huge, and covers a sea area of 766,309sq.km (295,874 square miles), or more than three times the land area of the country.

Fish in the waters around the British Isles are a wonderful resource, but fishing has to be responsible and sustainable. Concerns were raised as far back as 1885 that advances made in new technology was having a negative impact on both fishing stocks and their habitat. The fishing boom in the nineteenth century proved to be unsustainable and, after decades of overfishing, catches started to decline after the First World War; this trend has continued into the new millennium.

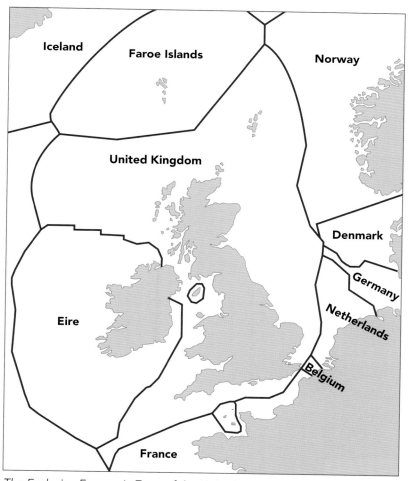

The Exclusive Economic Zone of the United Kingdom and neighbouring countries.

Wick was the herring capital of Europe. At its peak in 1867, it is claimed that 50 million herring were gutted and packed into barrels for export, in just two days. When the herring stocks collapsed, fishing moved on to white fish and cod, then more recently crab, lobster and scallops. But the industry here has never regained the heights of the Victorian age.

The number of regular and part-time fishermen has fallen from 47,000 in 1938 to 12,000 today, and the fishing industry is now less than 0.1 percent of the overall economy of the country. However, fishing still accounts for up to 40 percent of employment in some coastal communities, and the decline in fishing opportunities has devastated many of them.

After decades of decline there are some seeds of hope. The British government has promised further investment in the fishing industry and, following Brexit negotiations, British fishing quotas should increase in the future. This could herald the renaissance for British fishing communities, but only if lessons are learnt from the past and fishing in our coastal waters is performed responsibly and sustainably.

NAVIGATION

With more than two thousand years of coastal trade around the British islands, safe passage around our coastal waters has always been a priority for mariners. The Romans understood the importance of navigation and built two lighthouses at *Portus Dubris* (Dover) soon after their successful invasion in AD 43 – the one surviving structure is the oldest building in the United Kingdom (see photo on page 174).

During the Medieval period, local benefactors sometimes built lighthouses on hilltops near the coast, and they were often housed in religious buildings. On the Downs, near the most southerly part of the Isle of Wight, is a curious rocket-shaped building – Britain's only surviving Medieval lighthouse. St Catherine's Oratory is a tall octagonal stone tower, known locally as the 'pepper pot', and it once housed an internal light that acted as a beacon for ships passing south of St Catherine's Point.

The building was completed in 1328 and paid for by a local landowner as a penance for having misappropriated casks of wine from a shipwreck. The wine happened to be designated for a monastery in France, which is probably why the Catholic church took the theft so seriously. A small chapel was once attached to the side of the building where prayers were offered for the souls of shipwrecked mariners. The chapel was demolished in the sixteenth century during the desolution of the monasteries, but the octagonal tower survived beacuse of its importantance as a seamark.

As Britain's navy grew in size and international trade increased during the expansion of the empire, it became essential to improve safety at sea and to warn ships of offshore hazards. In September 1707, for example, Sir Cloudesley's fleet of 27 ships was returning from Gibraltar having encountered bad weather and poor visibility most of the way. As they approached the Isles of Scilly, the ship's navigators were unsure of their position. At 8pm on the evening of 2 November, four ships struck rocks to the south-west of St Agnes island and were lost. Between 1,400 and 2,000 sailors drowned, making the incident one of the worst maritime disasters in British naval history.

The Admiral Von Tromp was a Scarborough fishing trawler that ran onto the rocks in Saltwick Bay, Yorkshire, in September 1976. The exact circumstances of the tragedy remain a mystery, but the wreck is a timely reminder of the need for lighthouses and good navigation.

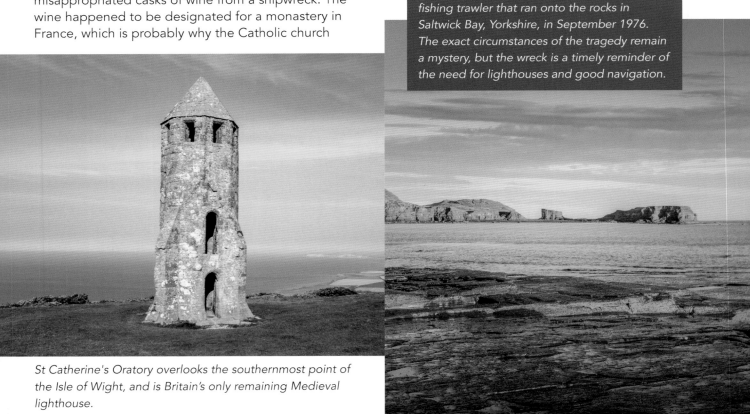

St Catherine's Oratory overlooks the southernmost point of the Isle of Wight, and is Britain's only remaining Medieval lighthouse.

Even before this catastrophe, work had already begun on the first lighthouse to be built in Britain in the modern age. The site was on the Eddystone Rocks, a dangerous reef about 19km (12 miles) south of the entrance to Plymouth Sound. Most of the reef is submerged at high water, which makes it particularly dangerous for ships heading towards the important harbour.

Construction on this rocky reef brought both success and disaster. The first attempt was a wooden structure built by Henry Winstanley and completed in 1698. Winstanley was an English painter and engineer and, during its construction, a French privateer took him prisoner and destroyed the foundations. This resulted in the French king, Louis XIV, claiming that '*France is at war with England, not with humanity*', and ordered Winstanley's immediate release. During the Great Storm of 1703, the lighthouse was completely swept away, together with Winstanley and five other men who were on the Eddystone attempting repairs.

A second lighthouse was soon started by John Rudyard, who was neither an architect nor an engineer, but a silk merchant and property developer. It was completed in 1709, and this time built of stone and lit with 24 candles. Despite Rudyard's apparent lack of proper qualifications, it lasted for 46 years before fire destroyed the structure. One of the three keepers died several days after being rescued having swallowed molten lead which had dripped from the lantern roof.

The third structure was erected by a civil engineer called John Smeaton, and it became a major step forward in lighthouse design. Built from Cornish granite blocks, the structure stood 18m (59ft) high and was operational by 1759. Smeaton pioneered the use of 'hydraulic lime', which was a concrete that cured underwater, allowing more substantial foundations to be built. The structure remained in service until 1877, by which time erosion had undermined the base, causing it to shake whenever struck by large waves. The lighthouse was dismantled stone by stone and rebuilt on Plymouth Hoe as a memorial; however, Smeaton's concrete foundations proved to be too strong to remove, and they remain today where they were laid in 1759.

Henry Winstanley's wooden lighthouse only lasted five years before being swept away in a storm, together with six men.

The second Eddystone lighthouse was much more substantial. Constructed of stone, it lasted for nearly half a century.

John Smeaton's lighthouse was operational from 1759 to 1877, before being moved to Plymouth Hoe.

The fourth and tallest lighthouse was completed in 1882 and is still used today, although it no longer houses lighthouse keepers. Smeaton's original foundations were too solid to dismantle, and they remain on the rocks today.

The fourth and final lighthouse on the Eddystone Rocks was completed in 1882. When it was built, the extra-tall lenses were the largest in the world. Although originally built to house several keepers, the beacon became the first offshore lighthouse to be fully automated in 1982, and a new helipad on the roof now allows safe and easy access for maintenance crews.

The last manned lighthouse in the United Kingdom was the North Foreland lighthouse in Kent, which became fully automated in 1998, ending exactly 300 years of the lonely occupation of a lighthouse keeper.

The lighthouse at South Stack on Anglesey was opened in 1809 and its main light can be seen for 44km (28 statute miles). The lighthouse was designed to assist the safe passage for ships on the treacherous Dublin–Holyhead–Liverpool sea route.

The Souter lighthouse in the village of Marsden in South Shields, Tyne & Wear, was opened in 1871 to warn vessels of the dangerous rocks of Whitburn Steel. In 1860 alone, there were 20 shipwrecks here, making this stretch of coastline the most dangerous in the country, with an average of more than 40 shipwrecks for every mile. This was the first lighthouse in the world to be designed and built specifically to use alternating electric current, and it was the most advanced lighthouse of its day.

Lighthouses are only part of a sophisticated navigational network that can be found throughout the coastal waters of the British Isles. These markers or buoys now conform to an international system of buoyage and range from the simple to the advanced. At their most basic, a stick poked into the mud (called a withy) can indicate the edge of a channel. Wooden or steel posts serve the same purpose, and they are often painted either red or green to mark the edge of the channel.

In British waters (and most other parts of the world), red markers and buoys indicate the left or port side of the channel as you approach from the sea; green markers and buoys indicate the right or starboard side of the channel. (Rather confusingly, the colours are reversed in the USA, South America and in the Philippines, Japan and Korea.) Sometimes these red and green channel buoys have a flashing light which corresponds to the colour of the buoy. The characteristics of any individual light is always marked on a marine chart, and this allows a navigator to identify a specific buoy.

A stick or withy is a simple but valuable method of marking the end of a muddy channel.

The entrance to Wootton Creek on the Isle of Wight is marked by conical port and starboard buoys at the entrance.

The other system of buoyage frequently seen around the coastline is a mixture of yellow and black buoys, referred to as Cardinal Marks. These are used to show the direction of the safest navigable water away from a mark. They might be used to mark the position of a wreck, shallow water or some other hazard. If, for example, a boat approached a South Cardinal, then it should steer south of the buoy to stay in safe water. These marks also have distinctive black triangles at the top which helps identification in difficult light conditions; at night, lights on Cardinal Marks are programmed with distinct characters which also helps to identify them.

VQ = Very quick

Q = Quick

(Number) = Flashes

LFl = Long flashes

s = Seconds

CLEAR TO NORTH
Black, Yellow
VQ or Q

DANGER

CLEAR TO WEST
Yellow, Black, Yellow
VQ(9) 10s or Q(9) 15s

CLEAR TO EAST
Black, Yellow, Black
VQ(3) 5s or Q(3) 10s

CLEAR TO SOUTH
Yellow, Black
VQ(6) + LFL 10s or Q(6) + LFl 15s

Cardinal Marks come in a variety of shapes and sizes, but their distinctive pattern of black and yellow banding clearly indicates their position relative to a hazard. The black triangles and sequence of flashing lights also assists in identification.

The network of buoys and lighthouses around the British Isles is maintained by Trinity House, a corporation established in 1514 under Henry VIII. Since 1998, the 60 lighthouses and offshore towers around the United Kingdom have been fully automated, but offshore buoys need regular maintenance and are checked every year, when they are repaired, painted, lifted and replaced when necessary. Even in an age of highly accurate Global Positioning System (GPS) receivers, electronic depth sounders, and sophisticated radar, it is essential to maintain these offshore navigational aids. Any technology can fail unexpectedly, and ships' batteries can run flat, so you always need a back-up system of charts and almanacs. And any fisherman or sailor will tell you that it is always reassuring to see a navigation buoy appear just where you expect it, or glimpse the loom of a lighthouse on the distant horizon on a dark night.

Buoys come in a variety of shapes and colours, each communicating something to the mariner. The major buoys and lighthouses are maintained by Trinity House.

CHAPTER 13

THE LEISURE COAST

The British coastline offers a huge variety of opportunities for watersports for all ages and all interests.

The British seaside means something different to each one of us: bucket-and-spade beaches, windswept dunes, craggy coastal paths or an invigorating sail in a brisk breeze. Whatever your fancy, you cannot stroll along a seaside promenade or walk across a sandy beach and not feel better with the world. The smell of the bracing sea air and the infinite timelessness of the ocean come together to invigorate the soul. But this feel-good factor is not all in the mind. Recent scientific research shows that time spent by the sea not only makes you *feel* better, but it actually *makes* you better. It encourages healthier lifestyles, lowers stress levels and could also stimulate your immune system – and the good news is that you can benefit from this even with a day trip to the beach.

THE HEALTHY COAST

It has been found that 'surf-generated aerosols' – essentially fine spray particles of seawater in the air – contain vitamins, salt, iodine, magnesium and trace elements which all have a positive effect on the body. These aerosols can stimulate an immune reaction on your skin and in your lungs, easing coughing and helping to clear airways. Recent research has also discovered that the tiny seawater droplets we breathe when near the coast could have a significant role in the prevention of lung cancer, and also on the level of cholesterol in our bodies – potentially leading to new preventative remedies.

Other research has shown that sea air is full of negative ions; these are charged particles which are abundant in sea spray, and which could improve our ability to absorb oxygen. Some studies suggest this might help us to feel more alert and could also help to decrease irritation from particles that make us sneeze, cough or have a throat irritation. Studies have also shown that inhaling salty air improves lung function in people with cystic fibrosis. Hay fever sufferers also benefit, because onshore breezes bring air from far out to sea which is free from pollen, exhaust fumes and soot particles.

Some doctors recommend 'coastal climate therapy' for people suffering with asthma, allergies or skin problems. The healing powers of a bracing climate are particularly effective in autumn and winter, when people with respiratory problems are most affected by the cold, air pollution or dry air caused by central heating. Getting into the water also has curative benefits, and magnesium in seawater may help improve skin condition, making it less dry and rough; this could help people with eczema or dermatitis.

At the University of Exeter in Devon, researchers are looking at the positive impact of the natural landscape on our wellbeing. They have found that a mix of green and blue colours is the most beneficial, which supports their findings that the closer people live to the coast, the healthier they say they feel. Psychologists believe this comes from reduced levels of stress, and an increase in physical activity from living near the sea. The research is also looking at how the benefits of a watery environment might be introduced into a clinical setting. By using virtual reality headsets with beach scenes, they hope to reduce a patient's pain and anxiety during, for example, dental treatment.

Time spent on the coast can also help your mental well-being. Psychologists have found that the natural environment is full of what they call 'soft fascinations', such as gazing at clouds or watching waves breaking on a beach. They call this 'attention restoration', and evidence shows that time spent in this state of contemplation helps replenish mental reserves, and that people actually concentrate better after spending time in the natural environment – or even simply by looking at scenes of nature.

THE SEASIDE RESORT

The main things that we associate with the British seaside, such as promenades and pleasure piers, all have their roots in the Victorian summer holiday – and the Victorians had all the right ideas, but for the wrong reasons. During Queen Victoria's reign, doctors increasingly recommended trips to seaside resorts for the bracing sea air, which contained what they called 'activated oxygen'. Physicians claimed this was *very essential but also a preventative of disease and a great aid for the treatment of ailments of all character.'* We now know there is no such thing as 'activated oxygen', but 'surf-generated aerosols' certainly do work, so Victorian doctors were not too far from the truth.

The promenade in Blackpool, c.1898. The electric tramline was the first in the country and opened in 1885. It only added to Blackpool's reputation as an exciting resort with state-of-the-art facilities.

A healthy stroll along the seafront was considered to be good for the Victorian constitution, but a walk along the esplanade was also somewhere to 'be seen' in polite society, where you could enjoy the glances of admirers as you strolled in your Sunday best. It became a popular pastime, and by the middle of the nineteenth century, seaside resorts began to grow rapidly.

The expansion of the railways transformed small coastal communities – often nothing more than small coastal villages – into bustling seaside resorts. Nowhere was this more apparent than in Blackpool, where its close proximity to the towns of the industrial north saw its popularity grow. In 1801, Blackpool was only a tiny fishing hamlet with fewer than 500 residents; by the end of the century, the population had grown to more than 45,000.

A railway branch line brought trippers from 1846 and the resort began to blossom. The promenade was built between 1856 and 1870, North Pier opened in 1863, then Central Pier in 1868. The Raikes Hall Garden opened in 1872, and the Winter Gardens in 1876. Electric lights were introduced in 1879, and the novelty attracted even more visitors. The famous tower was built between 1891 and 1894, confirming Blackpool's status as the top seaside resort in the north-west.

The Winter Gardens complex in Blackpool included an open-air roller skating rink. In the background is the Giant Ferris Wheel, which opened in July 1895, and quickly became a mecca for visitors.

In the early years, Victorian resorts were sedate and rather genteel places, with women wearing hats and crinoline dresses, even on the beach. By the end of the century, seaside towns had expanded not only in size but also in the attractions on offer. The new-fangled steam trains brought more tourists, and the sexes mixed openly on the promenade, bawdy 'what the butler saw' machines appeared in the arcades, and comedians told riské jokes in the music halls. Outside, the entertainment was just as raucous, with steam-driven fairground rides becoming increasingly popular.

The beach at East Parade, Bognor Regis, West Sussex c.1890. The bathing machines on the foreshore allowed people to change into swimwear and wade into the water, whilst still preserving their modesty.

As competition grew between resorts, towns became increasingly more innovative in what they offered the day tripper, including zoos, opera houses, theatres and aquaria. Early in the nineteenth century, the first seaside piers appeared, and were simple wooden jetties used to disembark passengers from steamboat trips. Ryde on the Isle of Wight has the world's oldest seaside pier, which opened in 1814 to receive trippers from the mainland ferries. Once the railway reached Southampton and Gosport on the mainland in mid-century, ordinary working people could travel to the island and visit the burgeoning resorts of Shanklin, Sandown and Ventnor. Ryde pier became an essential part of the tourist infrastructure, and the originally wooden construction was replaced with cast iron supports. Today, the Island Line railway still runs to the end of the pier to connect to the fast catamaran ferry service.

The Industrial Revolution brought new materials and new technologies; cast-iron pilings were introduced which were screwed into the seabed to act as supports. Piers became bigger and more sophisticated, offering a much wider range of entertainment and attractions, from theatres to penny arcades, and ballrooms to bowling alleys. By 1900, there were 80 pleasure piers in Britain, with some seaside towns having two or even three piers. Over the years many have deteriorated or been lost to fire, but 56 survive. In most cases the delicate ironwork, exotic lighting and ornate pavilions remain, and the piers still offer tourists the chance to take a promenade out to sea, even at low tide – all elegant relics from a bygone era.

The biggest pier in the country is at Southend-on-Sea in Essex, and it now stretches 2.16km (1.34 miles) out into the Thames estuary. British piers have always had a patchy record, with great popularity often being shattered by a variety of disasters – most notably fire and ship collision – and the history of Southend pier reflects that of many others throughout the country.

The first pier at Southend was completed in 1830, built entirely of wood and just 180m (600ft) long. It was soon found to be too short to be used by pleasure steamers at low water, and by 1848 it grew substantially to 2.1km (1.3 miles), making it the longest pier in Europe. By 1850, the railway from London reached the town and soon the resort was teeming with visitors. The wooden pilings deteriorated, and by 1889 a new iron pier was completed. It became an instant success, and within eight years an extension was built, followed by an upper deck added in 1907; the pier was extended yet again in 1927 to accommodate ever larger steamboats.

Ryde pier is still open to rail, cars, bicycles and pedestrians. The Isle of Wight railway uses an eclectic mix of London tube trains dating back to the 1920s.

Shortly before the Second World War, the pier was requisitioned by the Royal Navy, renamed HMS *Leigh* and closed to civilians in September 1939. It reopened to the public in 1945, and the number of tourists was greater than ever – and visitors only increased when a café, theatre and a hall of mirrors was opened. In 1959, a fire destroyed the pavilion near the shore, trapping more than 500 people on the pier, and they had to be rescued by boat. A second blaze in 1976 destroyed much of the pierhead, and a third fire the following year damaged the new bowling alley.

The pier at Southend-on-Sea after the old wooden structure had been replaced with iron in 1889.

Southend pier lives on, although now lacking some of its original gentrified charm.

The last fire was so destructive that the pier was closed in 1980, but public protest led to a substantial grant to make repairs. It opened again in 1986 only to have a ship crash into it, severing the pierhead from the rest of the structure. A further fire in 1995 destroyed the bowling alley yet again, and another blaze in 2005 badly damaged much of the old pierhead, including the railway station and a pub. A new pavilion at the end of the pier was opened in 2012 and includes a theatre. The pier has now been given Grade II listed status and lives on for the enjoyment of a new generation of visitors to the seaside resort.

The Victorians left us a legacy of the seaside holiday and we have them to thank that even today, we still all love to be beside the seaside.

The Grand Pier at Weston-Super-Mare opened in 1904, one of two piers in this small seaside town. It was destroyed twice by fire – once in 1930 and again in 2008. The pier won the National Pier of the Year award in 2001 and again in 2011, to become the first in Britain to win the award twice.

Birnbeck Pier at Weston-Super-Mare in Somerset, is emblematic of the sad decline of British coastal resorts. Built in the 1860s, it became a popular Victorian and Edwardian seaside attraction, and a boarding point for Bristol Channel steamers. As the popularity of British holidays declined, so it became financially unsustainable and was closed in 1979. The pier has since changed ownership several times and remains in a dangerous and derelict state, even though it is listed as a Grade II* building.

There are dozens of funiculars or cliff railways still working in our coastal towns, offering convenience and novelty. The first was the Scarborough South Cliff Tramway Company Limited, built in 1873. Like most early funiculars, two counter-balanced coaches were linked by cables, and water was used to alter the weight, causing one carriage to rise and the other to fall. Most have since been converted to electricity.

One of the new features of the great British seaside holiday in the early twentieth century was affordable holiday camps, which grew up between the two world wars. These camps came as complete packages, with food and family entertainment provided for the equivalent of an average man's weekly pay, and they replaced mundane boarding houses with a seaside fantasy resort in miniature. William 'Billy' Butlin opened his first site in Skegness in 1936 and, although he did not invent the holiday camp, he refined the concept and took it to a new level. Others copied the idea, and by 1939 there were several hundred holiday camps throughout the country, with the smallest housing 50 people and the largest accommodating more than 6,000 visitors.

Travel to holiday camps was mostly by coach, and campers were greeted by the entertainment staff on arrival – blue coats for Pontins and red coats for Butlins. Three meals a day were served in a communal dining hall, and entertainment for every age ran all day and into the evening. Most importantly, all activities, including the cinema and fairground rides, were part of the package, making the holiday ideally suited for families on a tight budget.

Six years of war had resulted in the British coastline no longer being accessible to the public, but from 1945 the seaside resort had a resurgence – partly because of rising living standards, and also because workers were guaranteed an annual vacation thanks to the Holiday Pay Act of 1938. During the 1950s and 1960s, few people could afford to go abroad, but took their holidays in Britain. Those living in the industrial towns of Manchester, Liverpool and Glasgow would most likely go to Blackpool or Morecambe; if you were from Leeds, you might choose Scarborough or Filey; and Londoners would prefer Brighton, Margate, Southend-on-Sea or possibly the West Country.

Despite the popularity of the holiday camp, the traditional seaside resort still flourished with amusement arcades, candyfloss stalls and Formica and chrome cafés serving fish and chips with mugs of steaming tea. Donkey rides, crazy golf and the inevitable dodgems were all on offer. Set back from the beach were neatly trimmed public gardens with a

Beach huts at Dawlish Warren, Devon.

bandstand and deck chairs. Many resorts also offered brightly painted beach huts which could be rented by the day or the week – useful for changing, or as a refuge from inclement weather. The resorts of the 1950s and 60s certainly looked very different from their Victorian origins.

The great era of British seaside holidays came to an end with the arrival of cheap foreign package tours, initially to Spain and then further afield, where the hot sun shone, and the cheap alcohol flowed. Straw sombreros and plastic flamenco dolls replaced sticks of seaside rock and seashells. Back at home, some holiday camps diversified into providing static caravan accommodation, but generally British seaside resorts took a turn for the worse, and a lack of year-round jobs and poor transport only adding to their decline. Some resorts hosted cheap temporary accommodation for the homeless, or for migrants seeking asylum, and this influx often caused resentment among some of the permanent population.

Some resorts did little to help themselves. In the 1980s, many of the country's beaches were frankly disgusting, and visitors found the idea of wading through a soup of litter and sewage less than appealing. The British government even tried to claim that beaches in popular resorts such as Blackpool, Brighton and Skegness were not actually bathing beaches – this way they tried to avoid having to deal with raw sewage flowing into the sea in these towns. Of the 27 beaches the government agreed

were actually used for swimming, nine were too dirty to reach the minimum bathing standard set by the European Union (EU).

It was this EU legislation which forced the clean-up of Britain's coastline, but the improvements did not happen overnight. Even by 1995, over half of England's 370 tourist beaches were fouled by unacceptable levels of pollution, but things have gradually improved. Even though the environmental regulations today are even tighter than they were back in the 1980s, there is no longer any need for caution. With 99 percent of Britain's designated beaches now considered safe for swimming, it is time to go back into the water.

The tide also seems to have turned for the British seaside resort, and the last couple of decades have seen a striking revival in their fortunes. The renaissance started with southern towns such as Margate, Hastings and Whitstable, and northern towns such as Whitley Bay in North Tyneside have followed the idea and began their own renewal after decades of neglect.

Some resorts diversified and used hotels and entertainment facilities to open conference centres. Others have invested in becoming major art venues – the Tate Gallery in St Ives being a particular success. In Scarborough, the Stephen Joseph Theatre is a centre for all of Alan Ayckborn's popular plays, and this has helped make the town Yorkshire's most successful resort.

A familiar sight around the British coastline. By the mid-1960s, one in seven British holidays was in a caravan and parks appeared in large numbers from the 1970s, offering cheaper self-catering family vacations as an alternative to foreign package deals.

Other new trends in recent years include mini-breaks, 'staycations' and Airbnb, all of which have contributed to a regeneration of the British seaside resort.

GETTING WET

Increased leisure time and rising incomes in post-war Britain have allowed more and more people to enjoy the British coastline, either on, around or under the water. In 2016, 4.7 million people (that is nearly 8 percent of UK adults) enjoyed some type of watersport, whether sailing, kayaking, surfing or paddleboarding. This is a huge increase in the last 30 or 40 years.

The coastline around the British Isles is there to be understood, respected and enjoyed, and boating is no longer something for the wealthy. You can easily spend half a million pounds on a new 12m (40ft) yacht, but you can also pick up a second-hand kayak or sailing dinghy for a few hundred pounds, which will give you endless enjoyment.

It is now easier than ever to learn new skills on the water. If you are interested in staying above the water, there are many places offering courses certified by the Royal Yachting Association (RYA). They offer training in sailing, powerboats and windsurfing; and you can also do theory courses over the winter months, including online.

Windsurfing goes back to 1948, but the sport boomed in the late 1970s. It became an Olympic sport in 1984 for men and in 1992 for women.

Other watersports such as kayaking, paddleboarding or surfing offer a cheap and easy introduction to the thrill of being on the water, and the equipment can often be hired at holiday resorts. You do not need formal training for any of these sports, but it can add to your enjoyment and decrease the time it takes to become competent. Courses endorsed by British Canoeing, BSUPA, International Surfing Association and Surfing England are recommended. In addition, by joining a club, you will meet people with similar interests and pick-up useful advice on the way. A quick online search will tell you where to go and who to contact for whichever watersport takes your fancy.

Sea kayaks and lifejackets can often be hired by the hour or the day, making it an ideal way of seeing how much you enjoy the experience. Inflatable kayaks now offer an alternative with easier storage and transportation.

Stand Up Paddleboarding is the fastest-growing watersport. (Taken from Stand Up Paddleboarding: A Beginner's Guide)

No need to head for Malibu or Morocco to catch a breaking wave. Thurso East in Scotland (shown here) and Freshwater West in Pembrokeshire offer some of the best surfing beaches in Europe for expert surfers. (Taken from Ultimate Surfing Adventures)

Kiteboarding and kitesurfing are different names for an extreme sport that is becoming increasingly popular, where you are pulled across the surface on a board by a 'traction kite'.

Four control lines run from the kite to a control bar, which allows the surfer to steer. A kite is much larger than a windsurf sail, so it will produce power in lighter winds; as the strength of the wind increases, smaller kites are used.

Open water swimming has become increasingly popular over the years, and it is obviously something you can do alone, providing you take appropriate precautions. There are also many events organised around the country for all abilities, which allow you to join like-minded swimmers – and with the support of safety boats. The bold and adventurous can swim the English Channel in a relay, or the less experienced can take advantage of an organised swim across the Solent from the mainland to the Isle of Wight. The Scilly Swim Challenge starts at sunrise from the harbour at St Mary's, and participants face the challenge of swimming between all six of the main islands that make up the Isles of Scilly. Many of these open swimming events take place in September when the sea is at its warmest, and you can still (hopefully) rely on some decent weather.

If you are more interested in going under the waves, then a simple snorkel, mask and flippers will give you endless fun and an insight into a fascinating underwater world; you can spend hours swimming safely in shallow water but be aware that your back can easily get sunburnt. Snorkeling is a great introduction to more serious diving using self-contained underwater breathing apparatus (scuba), but you need to take training seriously, first in a pool and later in open water. This instruction will include how to use the equipment, how to deal with the general hazards underwater, emergency procedures for self-help, and how to give assistance to another diver who might be experiencing problems. The British Sub-Aqua Club (BSAC) offer courses throughout the country.

SAFETY

We have always been well served in the United Kingdom with professional organisations dedicated to maritime safety. Her Majesty's Coastguard (HMCG) is a section of the UK Maritime and Coastguard Agency which is responsible for the co-ordination of maritime search and rescue (SAR). The Coastguard is recognised as the fourth emergency service, and they can be contacted by dialing 999. HMCG will respond to anybody in distress at sea, or at risk of injury or death along the coastline. The Coastguard is also responsible for land-based search and rescue helicopter operations and relies on both volunteers as well as full-time officers.

It is important to take a PADI (Professional Association of Diving Instructors) course; then, with the right qualifications, the underwater world is your oyster.

During the 1990s, the Coastguard service was rationalised and modernised, and this resulted in the closure of most of their coastal visual watch stations. Tragically, two fishermen in Cornwall drowned close to a Coastguard station that had recently shut. In response, the National Coastwatch Institution (NCI) was founded in 1994 as a charity. The organisation now has more than 2,600 trained volunteers, manning 56 coastal stations. The NCI personnel play a valuable role in visually monitoring the coastline on a daily basis from 8am to dusk; in addition to reporting serious incidents at sea, they are involved with many hundreds of minor events including finding missing children, assisting with distressed marine wildlife, and reporting pollution and dangerous debris washed ashore. The NCI liaise closely with HMCG, HM Revenue and Customs, the Home Office Border Force and the Royal National Lifeboat Institution (RNLI).

The RNLI operates 444 lifeboats out of 238 lifeboat stations around the coastline of the United Kingdom and the Republic of Ireland. The service was founded in 1824 and was originally called the Institution for the Preservation of Life from Shipwreck. Since then, the RNLI has saved more than 140,000 lives, with more than 600 volunteers over the years making the ultimate sacrifice. The early lifeboats were powered by oars and sail and were narrow open-decked boats designed to be self-righting. From the early 1900s, the lifeboats were powered by engines and increased in size. One little-known part of its history is the role played by the RNLI during the Dunkirk evacuation in 1940, when 19 lifeboats crossed the English Channel; the *Prudential* from Ramsgate repatriated 2,800 troops from the beaches.

The biggest RNLI lifeboat now in operation is the 17.3m (57ft) long Severn class, which costs £2.6 million to build. The lifeboat operates with a crew of seven and has a range of 250 nautical miles (463km); it can accommodate up to 124 survivors. The RNLI operate five classes of all-weather lifeboats as well as smaller inflatable inshore lifeboats, and the organisation saves, on average, one life for every day of the year. The RNLI also operate an extensive education programme, saving more lives by offering safety advice to boat and beach users, anglers, divers, kayakers and school children.

The self-righting ability of a lifeboat is tested by tipping it over with a crane at the Royal National Lifeboat Institution's store yard beside the Limehouse Cut, around 1885.

In 1974, a postage stamp was issued to mark the 150th anniversary of the RNLI. It showed the rescue of the crew of the Daunt Lightship in 1936 by the Ballycotton lifeboat, Mary Stanford. The lifeboat crew spent 49 hours at sea (25 hours without food), to save the lives of the eight men on board the drifting lightship.

In addition to the RNLI, there are more than 80 independent lifeboat services operating throughout Britain and Ireland. These smaller services tend to be overshadowed by the RNLI, but they all offer a valuable service and often work in partnership with the Coastguard and the RNLI. A typical example is the Freshwater Lifeboat Station on the Isle of Wight, which operates two inflatable lifeboats, and covers the south-west coast of the island up to 48km (30 miles) offshore. The volunteer crew are on call to the Coastguard 24 hours a day, 365 days a year, and their operation is entirely funded locally through a mix of donations, local business sponsorship and community grants.

As well as their lifeboats, the RNLI also provide lifeguards around many of Britain's most popular bathing beaches.

Watersport Safety

Safety is paramount, regardless of which watersport you enjoy:

- Always let someone know where you are, what you are doing and when you expect to return.
- Always check the weather forecast and tides before you leave.
- Plan your exit strategy before you start, and be aware of currents, tidal flow and wind direction (avoid offshore winds).
- Look out for safety signs, online information and local feedback.
- You need to be experienced if you are not under the advice of an instructor, and do not exceed the limits of your ability.
- Be aware that weather, temperature and water conditions can quickly make things difficult.
- Wear an appropriate lifejacket or buoyancy aid for your sport and avoid drinking alcohol.
- Even in a British summer, conditions can be cold, so consider wearing a wetsuit. This will insulate you against the cold, keep you buoyant in the water, and protect against knocks and scrapes.
- If you get into difficulty in the water, float on your back until you can control your breathing. Most importantly, stay calm and do not panic.
- If necessary, adopt the 'Heat Escape Lessening Posture' (HELP): draw your knees to your chin, keep your legs together, press both arms against your side, and keep your head out of the water.
- If you need assistance in the water, signal for help by raising one arm above your head with an open hand and shout for help.
- If you have the equipment, use a VHF radio on channel 16 or a mobile phone by dialing 999 and ask for the Coastguard.
- Orange distress flares can summon help in daytime or use the international SOS distress signal at night of three short flashes, three long flashes, followed by three short flashes.

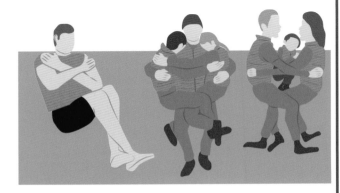

The HELP position is an invaluable cold water survivial technique which helps retain body heat. The huddle can be used to good effect when there is more than one person.

STAYING DRY

If you prefer to stay on dry land, there are dozens of organisations and charities dedicated to conservation. The National Trust, English Heritage, the Royal Society for the Protection of Birds (RSPB), the Wildlife Trusts, the Mammal Society, the World Wide Fund for Nature (WWF) and Whale and Dolphin Conservation (WDC) are just a few of the many charities who are actively conserving plants, animals and habitats around our coastline. There is even a charity for the conservation of seahorses – The Seahorse Trust. They all have excellent, informative websites to check out.

Walkers are also well catered for, and the British coastline offers some of the best scenery to be had anywhere in the world. The South West Coast Path, for example, is England's longest marked footpath and is classified as a National Trail. It runs for 1,014km (630 miles) from Minehead in Somerset, around the coastlines of Cornwall and Devon, to Poole harbour in Dorset.

The Wales Coast Path runs close to the entire coastline of Wales and is 1,400 km (870 miles) long. The walk was launched in 2012 and claims to be the first dedicated footpath anywhere in the world which covers the entire length of a country's coastline.

Scotland has dozens of fine coastal walks, including the Fife Coastal Path (188km or 117 miles) from Kincardine to Newburgh. Shorter walks include Portpatrick to Killantringan Lighthouse in Dumfries and Galloway, the Coffin Roads on Isle of Harris in the Outer Hebrides, and the spectacular walk from Cruden Bay to Bullers of Buchan in Aberdeenshire.

In Northern Ireland, the 53km (33 miles) walk from Portstewart to Ballycastle will take you through the Causeway Coast Area of Outstanding Natural Beauty.

By their very nature, coastal walks can vary from a gentle stroll along a sandy beach to a demanding and strenuous cliff walk, sometimes with very steep sections. There are plenty of guidebooks and websites available to help you choose the walk that will suit you best. VisitEngland, VisitScotland, Visit Wales and Tourism NI are all charged with promoting leisure activities in their respective nations, whether you are a rambler or prefer more sedentary pursuits. They all have good websites.

Beach safety does not always apply to just walkers and swimmers

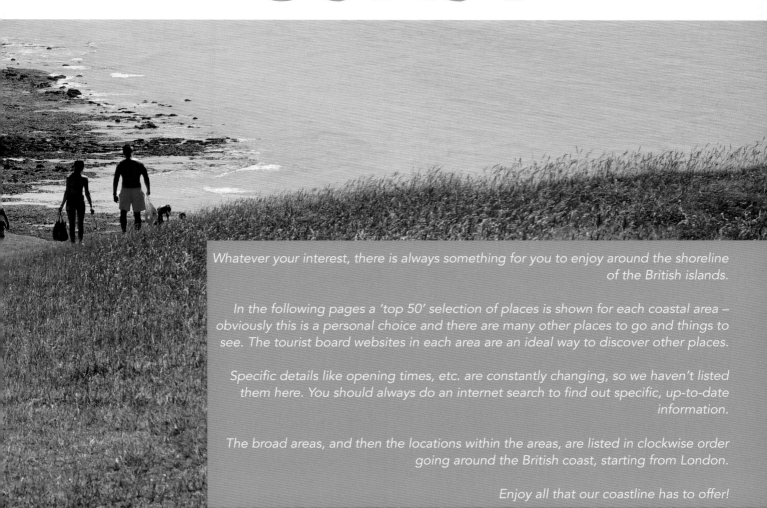

PART 4

DISCOVERING THE BRITISH COAST

Whatever your interest, there is always something for you to enjoy around the shoreline of the British islands.

In the following pages a 'top 50' selection of places is shown for each coastal area – obviously this is a personal choice and there are many other places to go and things to see. The tourist board websites in each area are an ideal way to discover other places.

Specific details like opening times, etc. are constantly changing, so we haven't listed them here. You should always do an internet search to find out specific, up-to-date information.

The broad areas, and then the locations within the areas, are listed in clockwise order going around the British coast, starting from London.

Enjoy all that our coastline has to offer!

LONDON, KENT & SUSSEX

The shape of the coastline of the British Isles results from the underlying bedrock, combined with the effects of waves, currents and the weather. In the south-east of England, alternating bands of rock of differing resistance run sometimes parallel to, and sometimes across the coast. This results in the Kent and Sussex shoreline alternating between sandy bays and chalk headlands. The Thames estuary is predominantly low-lying mudflats, with its own distinctive character and long history of trade.

Because of their proximity to the continent, Kent and Sussex are home to the best examples of Britain's coastline at war. From Roman and Medieval castles to more recent Martello towers and concrete pillboxes, this region is a great place to explore the history of Britain's sometimes cantankerous relationship with our near neighbours.

PHYSICAL COAST

Location	Features	Description
River Thames, London	Estuary, coastal plain estuary (ria)	One of the biggest estuaries in the country, extensive mud- and sandflats can be found on both north and south shores.
Isle of Sheppey, Kent	Mudflats, fossils	Fossilised turtle remains, lobsters, crabs, shark's teeth, snake remains and plants. Take care as the clay is very soft.
Beltinge (towards Reculver Towers), Herne Bay, Kent	Mudflats	London clay fossil remains include lobsters, crabs, shark's teeth, snakes, crocodiles and plant remains in Eocene rocks, 56 to 34 million years old.
Herne Bay, Kent	Beach, mixed shingle, sand at low tide	Shells can be found on sand on the lower foreshore; also look for sea glass, pebbles and fossils.
North Foreland, Kent	Cliffs, chalk, fossils	A prominent chalk headland commanding uninterrupted views across the southern North Sea; fossils can be found in the in chalk; flint pebbles.
White Cliffs, Dover, Kent	Cliffs, chalk	An iconic national landmark formed 70 million years ago, when skeletons of microscopic algae (coccoliths) were laid down in a warm, shallow sea.
Dover to Folkestone, Kent	Cliffs, fossils	Abundant fossils in the cliffs and foreshore around Samphire Hoe, Pegwell Bay, Kingsgate and St Margaret's Bay.
Dungeness, Kent	Beach, shingle, cuspate spit	The largest area of shingle in Europe which has evolved over 10,000 years since the last Ice Age. Pebbles and rare plants; RSPB nature reserve.
Camber Sands, East Sussex	Beaches	A large, gently sloping beach of fine sand and popular for swimming. However, there is a very real danger of quicksand.
Beachy Head, near Eastbourne, East Sussex	Cliffs, chalk, lighthouse	The highest chalk sea cliff in Britain, rising to 162m (531ft) above sea level and part of the South Downs National Park.
Seven Sisters, East Sussex	Cliffs	Striking vertical cliffs within South Downs National Park.
Seven Sisters & Birling Gap, East Sussex	Cliff erosion	Part of the South Downs National Park, the rolling landscape is caused by the remnants of dry valleys in the chalk hills (see picture pp 216-217).
Medmerry Beach, near Selsey, West Sussex	Beach, fine sand, dangerous	The Coastguard has warned the public to avoid Medmerry beach due to hidden dangers, including unexploded ordnance; the beach is now closed.
Bracklesham Bay, West Sussex	Beach, shingle becomes sandier towards the water's edge	You can find fossils washed up in the sand, including shark's teeth and turtle shells. Storms expose more fossils in winter and spring.

LIVING COAST

Location	Features	Description
Kent coastline	Marine Conservation Zones (MCZs)	Six MCZs were established in 2013; Medway estuary, Swale, Thanet coast, Dover to Deal, Dover to Folkestone, and Folkestone Pomerania.
Isle of Sheppey, Kent	Bird watching, raptors	The island has a high number of overwintering birds of prey; look out for the short-eared owl, harriers, merlin, peregrine falcon and rough-legged buzzard.
Elmley, Kent	Bird watching	The Isle of Elmley is part of the Isle of Sheppey, and the only Nature Reserve in Britain where you can spend the night; good for marsh harriers.
Swale Channel, Thames Estuary, Kent	Bird watching, waders and wildfowl	The Swale is a tidal channel separating the Isle of Sheppey from the rest of Kent; a 6,509-hectare (16,085-acre) Site of Special Scientific Interest (SSSI).
Lade Pits, Denge, Kent	Bird watching	Originally a 70 hectare (173 acre) gravel extraction site now run by the RSPB. See marsh harriers, egrets and the rare Sussex emerald moth.
Dungeness, Kent	National Nature Reserve (NNR) and Special Protection Area (SPA)	This huge area includes shingle, freshwater pits, wildflowers and wet grassland; home to 600 plants species – a third of all wild plants in the UK.
Sussex coastline	Marine Conservation Zones (MCZs)	Nine MCZs have been established covering 22 percent of the Sussex coastline, including Pagham harbour, Kingmere and Beachy Head.
Pagham Harbour, West Sussex	Conservation site	One of the few undeveloped stretches of the Sussex coast, with a sheltered inlet creating an important wetland site; see godwits and egrets.
Medmerry Beach, near Selsey, West Sussex	Conservation site	Offers 10km (6.2 miles) of walks and cycle rides through inter-tidal wildlife habitats. The shingle beach here is dangerous and closed.
Chichester Harbour, West Sussex	Beach	Common seals can be seen in the harbour and around Thorney Island.

HUMAN COAST

Location	Features	Description
Tower of London, London	Historic, Norman castle	Originally founded by the Romans, this is now the best-preserved Norman castle. It was still used during the First and Second World Wars.
Thames Estuary, London	Bridge, road	The Queen Elizabeth II bridge is the only fixed road crossing over the Thames east of the City, and is the busiest estuarine crossing in the country.
Grain Tower, Isle of Grain, Kent	Fort	Built in the mid-nineteenth century to protect the Sheerness and Chatham dockyards from a perceived French naval threat during the 1850s.
Medway Estuary, Kent	Coastal villages, 'marsh parishes'	The river offered access to the strategic naval establishments at Rochester and Chatham; some local villages became known as 'marsh parishes'.
Rochester, Kent	Historic, castle	The Norman keep here dates from 1127 and defends the strategic crossing of the River Medway; it is now an English Heritage building.
Chatham, Kent	Dockyard	The town grew up around the dockyard and military barracks; nineteenth century forts offered protection. The dockyard closed in 1984.
Thames Estuary, near Sheerness, Kent	Historic, wreck	The SS *Richard Montgomery* sank in 1944 full of munitions and remains a danger to shipping. The wreck is visible at low tide from Sheerness.
Reculver Castle, Herne Bay, Kent	Historic, castle	This was one of the first Roman shore forts to be built to defend against Saxon raids. Now much of the site has been lost to sea erosion.
Margate, Broadstairs & Ramsgate, Kent	Seaside resorts	Popular seaside resorts close to London. Dickens frequently visited Broadstairs, where there is a museum about the writer's life.
North Foreland Lighthouse, Kent	Lighthouse	The last manned lighthouse in the country was fully automated in 1998, ending exactly 300 years of the lonely occupation of a lighthouse keeper.
Ramsgate, Kent	Harbour	A seaside town and one of the largest marinas in the country. For many years, the port was a base for cross-channel ferries.
Richborough Castle, Sandwich, Kent	Historic, castle	Rutupiae or Portus Ritupis was founded by the Romans, and later became a Saxon shore fort. Sited amongst the low-lying east Kent marshes.
Deal Castle, Kent	Historic, castle	An artillery fort built by Henry VIII between 1539 and 1540 to defend the strategically important Downs anchorage off the Kent coast.
Walmer Castle, Kent	Historic, castle	Another Tudor artillery fort originally constructed between 1539 and 1540. The extensive ornamental gardens date from 1790.
South Foreland, St Margaret's Bay, Dover, Kent	Lighthouse	The first lighthouse to use electric light. The first ship-to-shore radio transmission and distress signals were also received by this lighthouse.
Dover, Kent	Historic, harbour	The Dour estuary at Dover formed a natural harbour here for the Roman town, but the river silted up during the Medieval period.
Dover Castle, Kent	Historic, castle	One of the 3 castles William of Normandy built after invading in 1066.
Dover Castle, Kent	Historic. lighthouse	One of two lighthouses built by the Romans, and one now survives in the grounds of Dover Castle. This is the oldest building in England.
Dover Harbour, Kent	Port	The country's main port of entry to and from Europe.
Sound Mirror, Abbot's Cliff, Kent	Historic, modern era, defence, sound mirror	These forerunners of radar were built between 1916 and the 1930s as 'listening ears', to provide early warning of incoming enemy aircraft.
Romney, Hythe and Dymchurch Railway, Kent	Railway	This narrow-gauge steam railway runs for 22km (13.6 miles) along coast and is one of the county's major tourist attractions; allow 1hr 10mins.
Acoustic Mirrors, Lade Pits, Denge, Kent	Historic, modern era, defence, sound mirror	Three concrete Sound Mirrors here range in size from 6m (20ft) to 60m (197ft). These Scheduled Ancient Monuments are now in a RSPB reserve.
Dungeness, Kent	Lighthouse	Site of seven lighthouses, with each new one built further out as the ness grows seaward. Now part of the RSPB Dungeness Nature Reserve.
Rye Castle, East Sussex	Historic, castle	Built by the Normans, the castle defended Rye – one of the Cinque Ports. Since used as a prison from the sixteenth to nineteenth centuries.
Pevensey Castle, East Sussex	Historic, castle	With 16 centuries of history, the castle chronicles the defence of the south coast during Roman, Saxon, Norman and Medieval times.
Beachy Head Lighthouse, East Sussex	Lighthouse	The 33m (108ft) lighthouse was built on the beach in 1903 to mark the headland, but the light is often obscured by sea mist.
Belle Tout Lighthouse, East Sussex	Lighthouse	A decommissioned lighthouse, now tea shop and B&B, with stunning views across the English Channel and the Beachy Head lighthouse.
Martello Tower, Seaford, East Sussex	Historic, nineteenth century fort	One of 103 towers built in the early nineteenth century as defence against a French invasion. Now a museum and one of the few open to the public.
Brighton & Hove, East Sussex	Seaside resort	From the 1730s, Brighton became popular for drinking and bathing in seawater as a cure for illnesses. The towns remain popular resorts.
Brighton, East Sussex	Pier	Grade II listed, opened in 1899.

ACTIVITIES

Location	Features	Description
Gravesend to Hastings, Kent	Walks & coastal paths	The Saxon Shore Way, beautiful diverse scenery; 246km (153 miles), classed as moderate, allow 13 days.
Ramsgate Marina, Ramsgate, Kent	Sailing, watersports	One of the largest marinas in the country, and several RYA Training Cen-tres are based here; Ramsgate Week is usually held in late July.

HAMPSHIRE, DORSET & ISLE OF WIGHT

The south coast of England offers some of the best weather in the country, and a variety of sandy beaches and traditional seaside resorts stretching from Rye, Hove and Brighton in the east, to Lyme Regis in the west. The Hampshire coastline comprises relatively young sands and clays, changing to more resistant and older limestones along the Dorset coast. There are many good, sandy beaches.

It is no accident that the Isle of Wight is called dinosaur island, and from the chalk deposits here to older limestones further west, this is fossil heaven. You can find anything from fossilised molluscs and wood, to crocodiles, turtles, fish, insects, plants and seeds. Further east in the limestones of Dorset, there are more opportunities to explore fossil-rich beaches around Lyme Bay, where crumbling cliffs regularly expose Jurassic fossils, including ammonites and even ichthyosaurs.

PHYSICAL COAST

Location	Features	Description
Bembridge Foreland, Isle of Wight	Fossils	Gastropods, brachiopods and other fossils are found to the east of the island.
Yaverland, Isle of Wight	Fossils	Famous for dinosaur remains (best in late winter at low spring tide).
Newtown Creek, Isle of Wight	Salt marsh and wetlands	This quiet backwater on the north coast was once a busy Medieval port; now, it is bustling with wildlife and the old town hall has no town.
Compton & other south coast bays, Isle of Wight	Fossils	Soft, easily eroded cliffs regularly expose fossils all along the south coast of the island; you can also find dinosaur remains and footprints.
Needles, Isle of Wight	Coastal features, sea stacks	An impressive row of three chalk sea stacks that rise about 30m (98ft) out of the sea off the western extremity of the Isle of Wight.
Southampton Water, Hampshire	Estuary, coastal plain	A large, drowned river valley which gives access to the port of Southampton, with its large cruise ship and container terminals.
Southampton Water, Hampshire	Tidal feature, double tide	Not caused by two entrances to the Solent as often supposed, but by the estuary's position halfway down the English Channel.
Hurst Spit, Hampshire	Coastal feature, barrier spit	A 1.5km (0.9 mile) shingle bank at the western end of the Solent, giving protection to salt marshes and mudflats.
Barton and Highcliffe, near Bournemouth, Dorset	Coastal erosion of clay and sand	Effective sea defences at Barton-on-Sea have limited the natural transport of beach sediment, resulting in very fast erosion of the coast to the east.
Hengistbury Head, Dorset	Coastal feature, barrier spit	A pebble beach partially blocking the entrance to Christchurch Bay; this barrier spit has moved and changed shape over the centuries.
Isle of Purbeck Peninsula, Dorset	Coastal features, cliffs, caves and wave-cut platform	At the eastern end of the Jurassic Coast are quarry workings, caverns and limestone ledges.
Chapman's Pool near Swanage, Dorset	Fossils	Rich in Upper Jurassic ammonites, reptile remains and shells. Part of the Jurassic World Heritage Coastline, so follow the fossil code of conduct.
Kimmeridge Bay, Dorset	Coastal features, wave-cut platform and shale cliffs	The bay lies within a marine Special Area of Conservation and has the best rock pooling and safest snorkelling in Dorset.
Jurassic Coast, Dorset	Fossils and limestone cliffs	150km (93 miles) of stunning, fossil-rich coastline running from the Old Harry Rocks, near Swanage to Orcombe Point near Exmouth, Devon.
Lulworth Cove, Dorset	Coastal feature, concordant coastline	The cove was created when the sea broke through a resistant band of limestone; this World Heritage Site attracts about 500,000 visitors a year.
Lulworth Cove, Dorset	Fossils	You can see 140-million-year-old fossilised trees along the cliffs just east of Lulworth Cove.
Old Harry Rocks, Dorset	Coastal features, sea stack, wave-cut notch, collapsed arch, cave	At Handfast Point at the southern end of Studland Bay, these features are some of the most famous coastal landmarks on the south coast.
Durdle Door, Dorset	Coastal feature, sea arch	This natural limestone arch near Lulworth rises about 60m (197ft) above sea level and is one of Dorset's most photographed and iconic landmarks.
Man O'War Ridge, St Oswald's Bay, Dorset	Coastal feature, concordant coastline	A line of pronounced and exposed rocks running in a straight line across Man O'War Cove results from the erosion along a concordant coastline.
Portland Bill, Dorset	Tidal feature, race and overfalls	One of the most notorious tidal races and overfalls in the British Isles; there is a good viewpoint of the race from the southern tip of Portland.
Chesil Beach, Dorset	Coastal feature, barrier beach, shingle	A shingle tombolo, 29km (18 miles) long, running from Abbotsbury to the Isle of Purbeck, encloses the saline Fleet Lagoon.
Charmouth, Dorset	Fossils	East of the Charmouth car park, the soft, low cliffs reveal plenty of ammonites, but do not use a hammer to remove them.
Church Cliffs, Dorset	Fossils	Large ammonites can be found along the foreshore but take care not to be cut off by a rising tide.
Lyme Regis, Dorset	Coastal features, wave-cut platform and rock pools	At low tide, an extensive rock ledge offers hundreds of rock pools to explore; the beach is covered at high tide.
Lyme Regis, Dorset	Fossils	An excellent site for ammonites.

LIVING COAST

Location	Features	Description
Poole Harbour, Dorset	Seals, common	Almost all sightings of common seal are in Poole harbour, but grey seals can be found all along the Dorset coastline.
Durlston Head, Dorset	Cetaceans	Dolphins can be seen from the cliffs.
Dorset coastline	Cetaceans and sharks	Approximately 10 cetacean species have been recorded here, including porpoises, various dolphin species, and minke and humpback whales.
Dorset coastline	Turtles	Leatherback turtles, the giants of the marine turtle world, can been seen during summer months in search of their favourite food – jellyfish.
Dorset coastline	Jellyfish and siphonophores	The coastline here is open to westerly winds from the Atlantic, so you frequently see many species of jellyfish and the Portuguese Man O'War.
Dorset coastline	Birdwatching	The sea cliffs all along the coastline give refuge and nesting sites to a very wide range of seabirds and raptors.
Dorset coastline	Seals, grey	Best viewed from headlands and bays in calm seas.

HUMAN COAST

Location	Features	Description
Ryde Pier, Isle of Wight	Pier	Oldest pier in Britain. Now the Portsmouth ferry terminus.
Ryde to Shanklin, Isle of Wight	Railway	The Island Line runs for 13.7km (8.5 miles) between Ryde Pier Head and Shanklin, using pre-war London tube trains; allow 50mins.
Coastal Artillery Battery, Culver Down, Isle of Wight	Historic, artillery fort	Overlooking the Eastern Solent and Nab Tower, this was one Palmerston's forts built on the island; served in both World Wars and closed in 1956.
St Catherine's Oratory, Isle of Wight	Historic, lighthouse	Curious rocket-shaped and Britain's only surviving Medieval lighthouse; its light acted as a beacon for ships passing south of the island.
Yarmouth Castle, Isle of Wight	Historic, artillery fort	An artillery fort from Henry VIII's frenzied castle building programme; built in 1547, it protected Yarmouth and the Solent from French incursions.
Spitbank Fort, Eastern Solent	Historic, artillery fort	The first of the 'Palmerston's Follies', built between 1861 and 1878. Declared a Scheduled Monument in 1967, it is now a boutique hotel.
Nab Tower, Solent	Historic, artillery fort and lighthouse	Built at the end of the First World War, it was placed in the Solent in 1920 at an angle of 3 degrees from the vertical, which can still be seen today.
Hayling Island, Hampshire	Seaside resort	The island combines a traditional seaside holiday with good sporting and leisure facilities, and excellent walks and cycle ways.
Portsmouth Harbour, Hampshire	Historic, fort, harbour, naval dockyard	Visit the Roman fort Portus Adurni, HMS *Victory*, the *Mary Rose* and many more, as well as seeing Royal Navy ships in the dockyard.
Mary Rose, Portsmouth	Historic, wreck	Discovered in May 1971 and raised in 1982. The preserved hull and many artefacts on board are exhibited in the Portsmouth Historic Dockyard.
Gosport, Hampshire	Naval base, sailing	Traditionally provided support for the Royal Navy across the water in Portsmouth, the town now has three large yacht marinas.
Southampton, Hampshire	Harbour, port	First operational in 1843, Southampton is now the busiest cruise terminal and second largest container port in the country.
Buckler's Hard, Hampshire	Historic shipbuilding, sailing	This nineteenth century village on the Beaulieu River was where ships for Nelson's Navy were built, using oak from the New Forest; now a museum.
Hengistbury Head, Christchurch Harbour, Dorset	Historic, harbour	Inhabited for at least 12,500 years; the harbour has been used since the Iron Age, although the entrance is now partially blocked by shingle.
Bournemouth, Dorset	Seaside resort	Bournemouth is currently voted the best beach in England, with miles of clean, golden sand, backdrop of cliffs, and plenty of little cafés.
Bournemouth, Dorset	Pier	A classic Victorian pier from 1880; after extensive refurbishment and rebuilding the thriving pier now boasts an IMAX theatre and climbing wall.
Poole Harbour, Dorset	Harbour	A busy cross-channel ferry port, despite the narrow entrance. Brownsea Island in the middle of the harbour has a population of red squirrels.
Swanage Pier, Dorset	Pier	A classic Victorian pier built for passenger ship services. It fell into disrepair in the 1960s but is open again with an aquarium and ferry service to Poole.
Weymouth Pier, Dorset	Pier	The pier was redeveloped for the 2012 sailing Olympic Games; it now has a theatre, pavilion, pleasure pier and a cross-channel ferry terminal.
Portland, Dorset	Historic, dockyard	Construction of the naval dockyard was completed in 1872; at the time it was the largest man-made harbour in the world.
Portland Castle, Dorset	Historic, artillery fort	One of Henry VIII's many artillery forts built between 1539 and 1541 as protection against invasion from France, and to defend Portland harbour
Lyme Regis, Dorset	Seaside resort	A genteel Georgian resort and fishing harbour, although the town dates back to the Domesday Book of 1086; home to fossil hunter Mary Anning.

ACTIVITIES

Location	Features	Description
Isle of Wight	Walks and coastal paths	The island's coastal path is 110km (68 mile) long, and a true walker's paradise; a network of other footpaths and trails criss-cross the island.
Freshwater Bay to the Needles, Isle of Wight	Walks and coastal paths, the Tennyson Trail	Best to walk west from Freshwater towards the Needles, and you will get great views of the Dorset coast and Old Harry Rocks ahead of you.
Compton & other south coast bays, Isle of Wight	Surfing and windsurfing	Good waves and stunning views across the Channel coast of the Isle of Wight.
Calshot, Hampshire	Sailing, windsurfing and paddleboarding	This small coastal village at the south-west corner of Southampton Water has one of the largest outdoor adventure centres in Britain.
Bournemouth, Dorset	Surfing reef	The Boscombe Surf Reef was an artificial reef built to enhance surfing conditions but proved to be a failure and closed after two years.
Kimmeridge Bay, Dorset	Surfing	Kimmeridge has three reef options for surfing; best conditions occur when a big south-westerly swell is combined with a north-easterly wind.
Weymouth/Portland Harbour, Dorset	Sailing and windsurfing	GBR's achievements in Olympic sailing and windsurfing are partly a result of the successful Weymouth and Portland National Sailing Academy here.

Alum Bay on the Isle of Wight, looking north-east towards Hurst Point and the lighthouse. The bay is well known for its multi-coloured sand cliffs which date from the Eocene, 60 to 34 million years ago.

Culver Down on the eastern side of the Isle of Wight overlooks the entrance to Portsmouth harbour. Because of its position, the headland was a military area for many years and closed to the public.

DEVON, CORNWALL & ISLES OF SCILLY

The West Country is a peninsula composed mainly of resistant limestones and sandstone, with some volcanic granite intrusions – all generally more than 300 million years old. When these rugged rocks are subjected to powerful wave erosion from the Atlantic, the result is a dramatic coastline of cliffs interspersed with sandy bays and coves.

These same rocks were an important resource in the past, and the north Cornwall coast especially has many examples of slate and mineral extraction, going back 4,000 years in the case of tin. Abandoned mines can be found all along the coastline, several of which have appeared in the *Poldark* TV series. The craggy coastline also provided good defence, and there are many examples of Iron Age 'cliff castles' to be explored, as well as Medieval castles, such as St Michael's Mount, Tintagel, St Mawes and Pendennis Castle.

PHYSICAL COAST

Location	Features	Description
Studland Bay to Exmouth, south Devon	Fossils, Jurassic	The Jurassic Coast is a UNESCO World Heritage Site where 185-million-year-old fossils can be found, but please following collecting guidelines.
Slapton, south Devon	Coastal feature, barrier beach	A freshwater lagoon has formed behind the bar.
Salcombe Estuary, south Devon	Estuary, coastal plain, offshore bar	The Salcombe-Kingsbridge Estuary is unusual because it has no large river feeding it; tidal up as far as Kingsbridge, and popular for sailing.
River Erme, south Devon	Estuary, coastal plain	The estuary is a classic drowned river valley. The river rises on Dartmoor and flows south to the coast; part of the South Devon Heritage Coast.
Fowey Estuary, south Cornwall	Estuary, tidal flats	Still a busy commercial river used by coastal vessels. The Fowey to Bodinnick ferry has been operational since at least the fourteenth century.
Carrick Roads, Falmouth, south Cornwall	Estuary, coastal plain, salt marshes and wetlands	A large flooded valley created after the Ice Age when sea level rose dramatically, which created a large, natural deep-water harbour.
Loe Bar, near Helston, south Cornwall	Coastal feature, barrier beach	A shingle barrier has blocked the River Cobber, creating a freshwater lake. Powerful waves and a slippery shingle bank make this beach dangerous.
Land's End, south Cornwall	Cliffs, igneous granite	A mass of granite rock formed about 280 million years ago is exposed at Dartmoor, Bodmin Moor, St Austell, Land's End and the Isles of Scilly.
Isles of Scilly	Coastal feature, tombolo	The islands of St Agnes and Gugh are joined by a sandy tombolo which is exposed at low tide and known locally as 'The Bar'.
Perranporth, north Cornwall	Beach, sandy	A huge, gently sloping sandy beach, perfect surfing, snorkelling, sailing and generally messing around in the water.
Penhale Sands, north Cornwall	Beach, sand dunes	Part of a Special Area of Conservation, and Cornwall's most diverse system of sand dunes. The South West Coastal Path runs through the dunes.
Bedruthan Steps, near Newquay, north Cornwall	Coastal features, sea stacks	Dramatic large slate rock stacks from the Middle Devonian period (386-377 million years ago) project up from a sandy beach.
Millook Haven, north Cornwall	Rock formations, sedimentary rocks	Sandstones and grey shales (marine mud) were deposited in an ancient sea 350 to 300 million years ago, with later tectonic folding.
Bude to Boscastle, north Cornwall	Fossils, Carboniferous fish and plants	There are many sites along this coastline where corals, brachiopods and goniatites can be found in limestone rocks along the foreshore.
Bude Bay, north Cornwall	Beach, sandy, coastal spit, quicksand	Bude claims to have more than 300 beaches which vary from iconic sand to intimate sheltered coves. Be aware of quicksand.
Crooklets Beach, Bude, north Cornwall	Beach, sandy bordered by rock pools	The sands around Bude contain calcium carbonate (a natural fertiliser), and in the past farmers took the beach sediment to spread on their fields.
Sharpnose Point, Morwenstow, Bude Bay, north Cornwall	Cliff erosion	Streams running into the bay have become truncated by wave erosion, leaving 30m (98ft) high coastal hanging valleys and waterfalls.
Hartland Point, north Devon	Cliffs	A chain of spectacular steep, rocky cliffs dropping almost vertically into the sea, interspersed with sheltered coves.
Westward Ho!, Bideford, north Devon	Beach, sandy, backed by a pebble bank	Fine sand and rock pools; inland is Northam Burrows Country Park, a grassy coastal plain with salt marsh, sand dunes, and grassland.
Braunton Burrows, north Devon	Sand dunes	The largest sand dune system in the British Isles, covering 13.6sq.km (5.2sq. miles). UNESCO World Biosphere Reserve and North Devon AONB.
Barricane Beach, near Woolacombe, north Devon	Beach, sandy, shells	A picturesque cove tucked in between the rocks, famous for cowries and other exotic seashells; folklore claims they come from Caribbean islands.
Woolacombe Beach, north Devon	Beach, sandy, backed by high dunes	Surfing, swimming, sailing.

LIVING COAST

Location	Features	Description
Berry Head, Brixham, south Devon	Cetaceans	Dolphins are regularly spotted offshore in calm waters; best viewed from a boat or headland.
Prawle Point, south Devon	Cetaceans	View from land or dolphin watching boats. The small minke whale is notoriously inquisitive around boats and might also be seen.
Isles of Scilly	Seals, grey	The islands are an Area of Outstanding Natural Beauty (AONB). The Atlantic grey seals here form a very important breeding population.
St Ives, north Cornwall	Seals, grey	Boat trips go out to Seal Island, just west of town, to see the seal colony.
Combe Martin to Morte Point, north Devon	Sharks and cetaceans, bird watching	Best views of basking sharks from a boat or headlands and bays when the sea is calm. This was Britain's first UNESCO World Biosphere Reserve.
Braunton Burrows, north Devon	Plants, wildflowers	This huge area of sand dunes has a rich diversity of vegetation and animal life, with 470 species of flowering plants.

HUMAN COAST

Location	Features	Description
Exeter to Teignmouth, south Devon	Railway	The train runs along the coast between Teignmouth and the Exe estuary, giving passengers fabulous views of the Devon shoreline; 25 minutes.
Kingswear Castle, south Devon	Historic, defence, Medieval	On the opposite side of the Dart estuary to Dartmouth Castle, built between 1491 and 1502 to complete the defence of Dartmouth.
Dart Harbour, Dartmouth, south Devon	Harbour	This natural, deep-water harbour can take cruise ships and naval vessels; it has been the home of the Britannia Royal Naval College since 1863.
Dartmouth Castle, south Devon	Historic, defence, Medieval	Built in response to the perceived threat from the French, the stone building dates from the 1380s.
Salcombe to Kingsbridge Estuary, south Devon	Harbour	Like other estuaries of south Devon, the deep river valley has been flooded by a later rise in sea level; a popular sailing destination.
Plymouth Sound, Plymouth, south Devon	Harbour	A large deep-water harbour and now home to the largest naval dockyard in western Europe.
Smeaton's Tower, Plymouth, south Devon	Lighthouse	Fully operational by 1759, but erosion eventually undermined its base. It was dismantled in 1877 and rebuilt on Plymouth Hoe as a memorial.
Eddystone Lighthouse, English Channel, south Devon	Lighthouse	The fourth and tallest lighthouse was completed in 1882 and is still in use today, although it is no longer manned.
Tamar River, south Devon and south Cornwall	Bridge, rail and road	The Royal Albert Bridge carries the railway line over the river. Designed by Brunel and opened in 1859. The new Tamar road bridge opened in 1961.
Charlestown, St Austell, south Cornwall	Harbour	A late Georgian working port which originally served the mining industry. It is now home to working tall ships used in TV and film period dramas.
St Mawes Castle, south Cornwall	Historic, defence, Medieval	An artillery fort built by Henry VIII between 1540 and 1542 to defend Falmouth against invasion from the French.
Pendennis Castle, south Cornwall	Historic, defence, Medieval and modern era	Built on a headland with breathtaking views out to sea, this Tudor castle was upgraded and played a vital role during the two world wars.
Lankidden, St Keverne, south Cornwall	Historic, defence, pre-history	A classic Iron Age 'cliff castle' enclosing more than a hectare of headland, reinforced with a double row of banks and ditches and a rampart.
St Ives to St Erth, north Cornwall	Railway	The railway line runs close to the shoreline around Carbis Bay before passing the Hayle estuary on its way to St Erth; 15 minutes.
St Ives and Hayle, north Cornwall	Harbour	St Ives is well known for its surf beaches and art scene, with the seafront Tate St Ives gallery offering rotating modern art exhibitions.
Winecove Point, St Merryn, north Cornwall	Historic, defence, pre-history	Three promontory forts south of Treyarno Bay, the southernmost of which has a large double ditch and bank which isolates the whole headland.
Padstow, north Cornwall	Harbour	The town is wrapped around a delightful stone fishing harbour; it is now one of Cornwall's top resorts made famous by a certain celebrity chef…
The Rumps, Minver, north Cornwall	Historic, defence, pre-history	Traces of round houses have been found on a rocky peninsula, and excavations here suggest the site was occupied for nearly three centuries.
Tintagel Castle, north Cornwall	Historic, defence, Medieval	A Roman fort later to become famous as the legendary birthplace of King Arthur; the existing Medieval castle dates from the thirteenth century.
Boscastle, north Cornwall	Harbour	The stone quay here is Elizabethan period and little has changed since. This tiny harbour was once one of the busiest in north Cornwall.
Hartland Quay, north Devon	Historic, defence, Medieval	The quay was built during Tudor times and was used allegedly by Sir Francis Drake to off-load the spoils of his privateering enterprises.
Ilfracombe, north Devon	Seaside resort	A popular seaside resort with a small harbour, good beaches, a world-famous aquarium, and the culinary and cultural capital of north Devon.
Lynton-Lynmouth, north Devon	Railway	The famous Lynton and Lynmouth funicular cliff railway opened in 1890, and is the highest and the steepest water powered railway in the world.

ACTIVITIES

Location	Features	Description
South West Coast Path, Dorset, Devon, Cornwall & Somerset	Walks and coastal paths	England's longest marked footpath is classified as a National Trail; it runs for 1,014km (630 miles) from Minehead in Somerset to Poole in Dorset.
Bigbury, south Devon	Surfing and other watersports	Bigbury Bay has a wide and varied selection of watersports available to visitors, including surfing, bodyboarding, kayaking and windsurfing.
Sennen Cove, north Cornwall	Surfing	Sennen Cove and Gwynver Beach offer some of the most consistent beach breaks in Cornwall, the bigger Gwynver waves are better suited to experts.
Perranporth, north Cornwall	Surfing	Offer a huge sandy beach with the best waves around low tide; be aware of rip currents when there is a big swell coming in.
Newquay, north Cornwall	Surfing	Towan Beach in the heart of town, and west-facing Fistral beach are the surfing hotspots here; there are several surfing schools and shops too.
Camel Estuary, north Cornwall	Walks and coastal paths	An old railway track which used to run along the edge of the estuary has now been turned into a walking and cycling path.
Camel Estuary, north Cornwall	Sailing and powerboats	An RYA training centre is based in the estuary, giving instruction in sailing dinghies, yachts and powerboats.
Woolacombe, north Devon	Surfing	Most of the surfing action is in the middle of town, but it is a long beach if you want to get away from the crowds.

Plymouth harbour looking south-east towards Renney Point in the distance. The 1.5km (1 mile) long stone breakwater protects shipping at anchor in Plymouth Sound and adjacent havens; it took 32 years to build and was completed in 1844.

SOMERSET, AVON & GLOUCESTERSHIRE

In contrast to the West Country, the southern shore of the Bristol Channel is primarily low-lying, with extensive mudflats as a result of a tidal range of up to 15m (49ft). Strong tides here create a lot of turbulence, and this gives the water a marked brown coloration. The tide goes out 1.5km (0.9 miles) or more, creating huge areas of salt marsh and mudflats; these attract an average 74,000 migrating birds each winter. Somerset is very good for fossils.

Although strictly in north Devon, a trip to the island of Lundy is a fabulous experience, although not cheap (a family day ticket was £98 in 2021). The graceful MS *Oldenburg* sails three times a week from Bideford or Ilfracombe, and the crossing takes about 2 hours each way. The island is only 5km (3.1 miles) long and less than 1km (0.6 miles) wide, and its 4,000 years of human history, listed buildings, natural rugged beauty and captivating wildlife make it a walker's paradise.

PHYSICAL COAST

Location	Features	Description
Bossington Beach, Somerset	Beach, barrier, shingle	This large, steep shingle barrier beach separates the Bristol Channel from the River Horner; a stream seeps through the shingle virtually unseen.
Minehead, Somerset	Beach, sand and shingle	Known local as 'The Strand', a wide expanse of sand with some shingle and many rock pools; good views from the top of nearby North Hill.
Blue Anchor, near Watchet, Somerset	Fossils	The cliffs here have a thin, hard Triassic bone bed full of fish and reptile remains; search the foreshore for these rocks and split them finely.
Doniford Bay, near Watchet, Somerset	Fossils	The foreshore is very good for white ammonites, as well as well-preserved brachiopods and bivalves.
St Audrie's Bay, near Watchet, Somerset	Coastal features, wave-cut platform, waterfall	This wide bay is a mix of pebbles, sand, shingle and rock. One of the nicest features here are two waterfalls which cascade down the cliffs.
Quantoxhead, Somerset	Cliff erosion, fossils	This pebble, rock and sand beach is backed by unstable cliffs revealing Jurassic fossils. At low tide a large wave-cut platform reveals rock pools.
Quantoxhead, Somerset	Fossils	The coast here has several kilometres of tall Jurassic cliffs and a long wave-cut platform; superb ammonites and reptile remains to be found.
Kilve Beach, Somerset	Fossils	Jurassic cliffs and foreshore; ichthyosaur and plesiosaur bones are the most common reptiles, also fish remains and various species of ammonite.
Lilstock, Somerset	Fossils	The beach here offers ammonites, shells and fish remains, and complete skeletons are regularly found.
Bridgewater Bay, Somerset	Mudflats, sand dunes, nuclear power station	The bay comprises mudflats, although winds have created sand dunes at Berrow; there are remains of a submerged forest dated 2,500-6,500 BC.
Stolford Beach, near Bridgewater, Somerset	Fossils	A large wave-cut platform of limestone and shale bands; the lower foreshore is very muddy, so look higher up for Jurassic fish and reptiles.
Sand Point Beach, Middle Hope, Somerset	Fossils	A wide, muddy, inter-tidal area has warning signs, so take care. Corals and crinoids can be found on the foreshore and in the cliffs.
Bristol Channel, west of England and south Wales	Navigation, estuary	The estuary can be a hazardous area of water because of strong tides, and few harbours accessible at low water; pilotage is essential for shipping.
Battery Point, Portishead, Somerset	Fossils	Devonian and Carboniferous rocks (about 420-300 million years old); corals, crinoids and fish can be collected from rocks on the foreshore.
Aust Cliff, Severn Estuary, Somerset	Fossils	A geological Site of Special Scientific Interest with reptile fossils from the Triassic period (250-200 million years ago) found in mudstone and limestone.
Severn Bore	Tidal feature	The narrowing of the Bristol Channel creates the world's second highest tides; the Severn bore creates a steep wave which attracts surfers.

LIVING COAST

Location	Features	Description
Somerset coastline	Cetaceans	Grey seals, harbour porpoises and occasionally dolphins can be seen from the shoreline.
Porlock Weir and Porlock Marsh, Somerset	Birdwatching	Rising sea level / flooding.
Hurlstone Point, near Porlock, Somerset	Cetaceans	Frequent sightings of harbour porpoises here, where they are believed to give birth and nurture their calves.
Bridgewater Bay, Somerset	Birdwatching, rare plants and insects	A Site of Special Scientific Interest (SSSI), it is particularly important for over-wintering waders and wildfowl, with 190 species recorded.
Gloucestershire coastline	Bird watching	The extensive mudflats which are exposed at low water provide excellent overwintering for waders and wildfowl.

HUMAN COAST

Location	Features	Description
West Somerset Railway	Railway, tourist	The longest heritage railway in England, running through the Quantock hills and along the Bristol Channel for 32km (20 miles); allow 1hr 20mins.
Watchet	Harbour	From 1564 onwards, the harbour here was used to import salt and wine from France, and export ore from the Brendon Hills; now a yacht marina.
Beach Beacon, Burnham-on-Sea, Somerset	Lighthouse	The wooden lighthouse on the beach at Burnham-on-Sea was built in 1832, and stayed in service until 1969; now a much-loved local landmark.
Berrow, near Burnham-on-Sea, Somerset	Historic, wrecks	The SS *Nornen* ran aground in a gale in March 1897, and the wreck can clearly be seen at low water, a stark reminder of the dangers to shipping.
Brent Knoll Iron Age fort, Somerset	Historic, Iron Age	Before the Somerset Levels were drained in the seventeenth century, Brent Knoll was an island, making it an easily defended Iron Age hill fort.
Brean Down forts, Somerset	Historic, pre-history to modern era	The current buildings date from the 1860s and were part of Palmerston's defence of the Bristol Channel ports; later used for weapons testing.
The Grand Pier, Weston-Super-Mare, Somerset	Pier	This pleasure pier was opened in 1904, but destroyed by fire in 1930 and again in 2008; it is now a listed building and officially re-opened in 2011.
Birnbeck Pier, Weston-Super-Mare, Somerset	Pier	Birnbeck opened in 1867 and became a popular boarding point for Bristol Channel steamers; it closed in 1973 and remains in a derelict state.
Worlebury Hill Fort, Sand Bay, Somerset	Historic, Iron Age	This Iron Age hill fort used earth walls, embankments and ditches to supplement the natural defensive position of a headland.
Clevedon Pier, Somerset	Pier	A long, elegant cast iron Victorian pier; the poet Sir John Betjeman declared it was 'the most beautiful pier in England'.
Battery Point, Portishead, Somerset	Historic, Medieval and modern era	Originally an Elizabethan watchtower, then Civil War battery, a Victorian coastal battery, and finally First and Second World War coastal defences.
Avonmouth	Harbour	Because of the high tidal range, the harbour is accessed via a 210m (690ft) long lock; the docks are important for food imports.
Bristol unitary authority	Harbour, historic	Bristol is a county in its own right, with a city centre harbour; a lock and docks were opened in 1809 to allow the use of bigger vessels.
SS *Great Britain*, Bristol	Historic, ships	Brunel's state-of-the-art steamship was launched in 1843, but scuttled in the Falkland Islands in 1937; now returned to Bristol and fully restored.
Severn Road Crossing, Aust, Somerset	Bridges, road	Opened in 1966, the bridge became a victim of its own success and a second crossing – the Prince of Wales Bridge – was opened in 1996.

ACTIVITIES

Location	Features	Description
South West Coastal Path	Walks and coastal paths	This 1,010km (630 mile) route runs from Minehead in Somerset to Poole in Dorset.
Minehead, Somerset	Windsurfing and kitesurfing	Minehead's north-facing sand and shingle beach was almost completely washed away in 1990; now good for swimming and watersports.
Watchet, Somerset	Surfing	The beach here is one of the best surf spots in the area, especially in winter.
The Severn Way	Walks and coastal paths	Starting in Bristol city centre, this long walk proceeds along the River Avon to Severn Beach, then up to Gloucester, Shrewsbury and into Wales.
The Jubilee Way	Walks and coastal paths	This 27km (17 mile) route links Aust at the south side of the Severn Bridge, and offers a scenic walk along the banks of the Severn estuary.

Porlock Bay in Somerset includes shingle ridges with salt marshes inland; offshore is a submerged forest. Much of the coastline here is in the care of the National Trust.

SOUTH WALES

This is a very diverse coastline, ranging from the muddy Usk estuary at Newport and becoming progressively more rugged as you travel west. You can explore it along the Wales Coast Path, which starts in Chepstow and ends in Chester, after a 1,400km (870 miles) walk. The south Wales coast is another good place for fossil hunting, and the rocks here offer ammonites, gastropods, belemnites and brachiopods, as well as the remains of fish and reptiles. 500-million-year-old graptolites can also be found.

To the far west, the Pembrokeshire Coast National Park offers a wide variety of rock types and landforms. It is this special geodiversity that led the coastline to be classified as a National Park in 1952. 40 percent of the park is protected as Sites of Special Scientific Interest (SSSI), so please respect the guidelines for visitors.

PHYSICAL COAST

Location	Features	Description
Coastline between Penarth and Porthcawl, Glamorgan	Cliffs, Jurassic and Triassic fossils and foreshore	A wide range of fossil including ammonites, belemnites, brachiopods, as well as reptiles, fish and shark remains; also look for dinosaur footprints.
Penarth, Glamorgan	Fossils	This is the most popular fossil location in Wales; although it is becoming over-collected, but you should still come home with some good finds.
Llantwit Major, Glamorgan	Coastal features, caves at base of cliffs, fossils	Once a smuggling hotspot, local legend claims many caves are linked by secret tunnels; excellent for Jurassic fossils, giant brachiopods and gastropods.
Southerndown, Glamorgan	Beach and cliffs	A west-facing beach with sand at low tide; accessed by steep steps; the cliffs are hard, grey, often shelly, Carboniferous Limestone.
Gower Peninsula	Beach, sand	There are more than two dozen beaches and coves on the Gower; facing west, the water is warmed by the Gulf Stream.
Oxwich Bay, Gower	Beach, sandy and salt marsh	A long sandy beach backed by large dunes, woodland, and the Oxwich National Nature Reserve with plenty of birdlife.
Paviland Cave, Gower	Coastal feature, cave	Also known as the Goat's Hole, lies between Port Einon and Rhossili; used as a burial site for an Old Stone Age male, dated 29,000 years ago.
Rhossili Bay, Gower	Beach, sand, collapsing cliffs	A magnificent curving sand beach facing the Atlantic Ocean; there was partial collapse of the cliff face during a storm in 2014, so take care.
Whiteford Beach, Gower	Beach, fine sand and mud	Located on the north-western tip of Gower, this 3km (2 miles) expanse of beach is part of a National Trust nature reserve.
Burry Inlet, between the Gower Peninsula and Llanelli	Estuary and salt marsh	An area of inter-tidal sand- and mudflats 22sq.km (8.5sq. miles), and the largest continuous area of salt marsh in Wales.
Burry Port, west of Llanelli	Beach, sand and sand dunes, salt marsh	West of the town is Pembrey Burrows, a large area of dunes and marshland and the site of many shipwrecks, many caused by wreckers.
Pembrey, Carmarthenshire	Beach, sand	13km (8 miles) of golden sand and 200 hectares (500 acres) of woodland makes this a perfect family holiday destination.
Pembrokeshire coastline	Cliffs and foreshore, this is an SSSI but loose fossils may be collected	Try looking around Abereiddy Bay, Druidstone Haven, West Angle Bay and Marloes Sands for good examples of a wide variety of fossils.
St Govan's Head, Pembrokeshire	Coastal features, sea stacks and caves	The Irish monk, St Govan, apparently lived in a small cave in the sea cliff, and a small chapel was built here in the thirteenth century.
Green Bridge of Wales, Pembrokeshire	Coastal feature, sea stacks, sea arch	A magnificent natural sea arch formed in Carboniferous limestone; only accessed through the Castlemartin Firing Range, so check before visiting.
Milford Haven, Pembrokeshire	Estuary, drowned river valley (ria)	The largest estuary in Wales with a 320km (200 mile) coastline, and one of the deepest natural harbours in the world.
Abereiddy, Pembrokeshire	Fossils, especially graptolites	Abereiddy is an SSSI so hammering into the bedrock is strictly forbidden. However, you can collect fossils from the flaky shales on the beach.

LIVING COAST

Location	Features	Description
Kenfig Bay, Bridge End, Glamorgan	Wild plants	One of the most important dune areas in Wales, and a treasure trove of rare plants, including wild orchids, sea stock, dune pansy and moonwort.
Burry Inlet, between the Gower Peninsula and Llanelli	Bird watching	Expect to see oystercatchers, pintails, dunlin, curlew, grey plover and several duck species including the shoveler, wigeon and shelduck.
Milford Haven, Pembrokeshire	Bird watching and cetaceans	This huge estuary is sheltered from the Atlantic and offers very diverse habitats for a wide range of seabirds, as well as bottlenose dolphins.
Skomer and Skokholm Islands, Pembrokeshire	Bird watching	A great habitat for ground nesting birds of all kinds including puffins, and the largest breeding colony of the Manx shearwater in the world.
Skomer and Skokholm Islands, Pembrokeshire	Seals, grey	Only grey seals occur regularly along the Pembrokeshire coast; go in winter to see pups between October and January.
Cardigan Bay between New Quay and Cardigan	Cetaceans	You can see bottlenose dolphins and harbour porpoises from the coast or from local boat trips.
Cardigan Bay between New Quay and Cardigan	Seals, grey	Grey seals are resident here all year, but they give birth to their pups during the winter months.

HUMAN COAST

Location	Features	Description
Newport, Gwent	Harbour	The Town Dock opened in 1842, but improved railway links and the growth in overseas coal trade, led to further significant expansion.
Cardiff Bay, Glamorgan	Harbour	With the decline of the coal industry the old docklands fell into disuse; its renaissance has created Europe's biggest waterfront development.
Cardiff, Glamorgan	Harbour	Cardiff owes much of its history to the Industrial Revolution; during the 1830s, the docks grew dramatically as the iron and coal trade increased.
Nash Point Lighthouse, Glamorgan	Lighthouse	Guarding the Nash Sands (sandbanks in Bristol Channel), this iconic lighthouse is a nineteenth century, Grade II listed, historic building.
Porthcawl, Glamorgan	Seaside resort	Its annual Elvis Festival attracts Presley imitators from all around the world; it was one of the great British resorts, and is still King for many.
Swansea to Pembroke Dock	Railway	The line snakes through south Wales and stops at stations giving access to the Pembrokeshire Coast Path; it is a slow train, taking two hours.
Swansea Docks, Glamorgan	Harbour	By the mid-nineteenth century, Swansea docks was exporting 60 percent of the world's copper, as well as zinc, iron and coal.
Tenby, Pembrokeshire	Seaside resort	This thirteenth century town and harbour is a top Welsh resort and part of the 'Welsh Riviera'; explore its sandy shore and the ruins of Tenby Castle.
Tenby Castle, Pembrokeshire	Historic, castle	Just a small tower remains of Tenby Castle, perched on Castle Hill; it was built by the Normans in the twelfth century, but later extended.
Manorbier Castle, Tenby, Pembrokeshire	Historic, Medieval	With a strategic position overlooking Manorbier beach, this Norman castle was a film set for the BBC's *The Lion, The Witch and the Wardrobe*.
Caldey Island, Pembrokeshire	Religious site	One of Britain's holy islands, Cistercian monks here continue a tradition going back more than 1,500 years; the residents welcome day visitors.
Milford Haven, Pembrokeshire	Harbour	The huge, natural deep-water harbour is the country's largest energy port, and also handles cargo, shipping layover, freight and passenger services.
The Solva Gribin Iron Age fort, Pembrokeshire	Historic, pre-history	A fine example of an Iron Age promontory fort built on a narrow ridge, guarding the entrance to Solva harbour.
St Davids, Pembrokeshire	Seaside resort	The smallest city in Wales with a big history; explore everything from a twelfth century Norman cathedral to the Oriel Parc art gallery.
Fishguard, Pembrokeshire	Harbour, natural	Fishguard is a busy ferry port with services to Ireland, and served by the West Wales Line from Swansea, with connections to Cardiff and London.

ACTIVITIES

Location	Features	Description
Wales Coast Path	Walks and coastal paths	This runs for 1,400 km (870 miles) and claims to be the first footpath in the world to cover a country's entire coastline.
Abertaw to Porthcawl, Glamorgan	Walks and coastal paths	The Glamorgan Heritage Coast (part of the Wales Coast Path) is 23km (14 miles) long and is suitable for both walkers and cyclists.
Llantwit Major, Glamorgan	Surfing	Best for experienced surfers as submerged rocks, potholes and rip currents can create problems.
Porthcawl, Glamorgan	Surfing	A good beginner's beach, with a surf school and surfboard hire.
Gower Peninsula	Walks and coastal paths	The circular footpath around Gower is 63km (39 miles) long and takes four to five days; however, the peninsula is criss-crossed with shorter routes.
Caswell Bay, Gower	Surfing	This flat, sandy beach generally has small surf so it is ideal for beginners; the waves can get bigger in the winter months but it is fairly sheltered.
Rhossili Beach, Gower	Surfing	Very good surfing; in 2013, Rhossili was voted the third best European beach by TripAdvisor, and the highest placed UK beach on the list.
Llangennith, Gower	Surfing	This long, south-westerly-facing beach offers consistency, but be aware of rip currents; best for expert surfers.
Manorbier, Pembrokeshire	Surfing	A fairly exposed beach with fairly consistent surf, although summer tends to be flat. Best around high tide, but beware of rip currents and rocks.
Freshwater West, Pembrokeshire	Surfing	The south-westerly-facing beach has consistent swell, and is considered one of the best surfing spots in Wales; best for experienced surfers.
Dale, Milford Haven, Pembrokeshire	Sailing / windsurfing	Dale is a popular watersport centre, giving access to well-protected sailing areas within the Milford Haven estuary.
Broad Haven, St Bride's Bay, Pembrokeshire	Sailing / windsurfing	A more exposed watersports centre facing west into the Irish Sea, with sailing centres and tourist boats to inshore islands.
Newgale, Pembrokeshire	Surfing	A huge sandy beach, a surfing school and board hire make this a good place for beginners; easily accessible.

NORTH WALES

This region too has a very varied coastline. The beautiful sandy beaches around the Barmouth estuary run almost uninterrupted to the north and west for 80km (50 miles). The coast faces the Atlantic, so it is not surprising that the area attracts surfers. However, the region is also well-suited to family holidays with young children.

In the north-west, the Llŷn peninsula has a rugged beauty, with sandy bays punctuated by rocky headlands. Further up the coast between Caernarfon and Bangor is the Menai Strait, separating Anglesey from the mainland. This narrow channel was carved by ice during the last glacial period. Between the two bridges is an area called the Swellies, where rocks near the surface create overfalls and local whirlpools. Whilst you are here, Anglesey too is well worth a visit. Like all islands around the British Isles, it has its own distinctive history and culture, and a beautiful coastline.

PHYSICAL COAST

Location	Features	Description
Borth, Ceredigion	Beaches, sandy	A wide sandy beach plus an ancient, submerged forest of oak, pine, birch, willow and hazel is visible at low tide, dated to about 1,500 BC.
Tywyn, Gwynedd	Beaches, sandy, dunes	The name comes from the Welsh ('tywyn', meaning beach); seashore, sand dune; extensive dunes are still found to the north and south of the town.
Fairbourne Spit / Penrhyn Point, Cardigan Bay	Coastal feature, shingle spit	Fairbourne Spit (aka Penrhyn Point) is a shingle spit created by longshore drift at the mouth of the River Mawddach, opposite Barmouth.
Criccieth, Llŷn Peninsula	Beaches, sandy	A popular family seaside town; the eastern shore of the bay has a sandy beach with a shallow area for bathing.
Porthor, Llŷn Peninsula	Beaches, whistling sands	The name Porthor, or 'Whistling Sands', comes from the squeak produced by walking on the unusual, shaped sand particles in warm weather.
Menai Strait, Anglesey and Gwynedd	Tidal feature, whirlpools	The strait has a rough, rocky bottom in places, and the tidal current can run at more than 8kts (14.8km/hr), creating numerous short-lived whirlpools.
Bwa Gwyn, Rhoscolyn, Anglesey	Coastal feature, sea arch	A stunning natural arch (Bwa Gwyn means White Arch) of white quartzitic rocks, displaying fractures and weathering: you can access the top of the arch.
Porth Dafarch, Trearddur Bay, Anglesey	Beaches, sandy	A sandy beach and rock pools makes this bay ideal for children of all ages, and the more adventurous can explore the rocky headland.
Traeth Mawr, Cemaes Bay, Anglesey	Beaches, sandy	A pretty village with two sandy beaches; at low tide you can explore sea caves and sandy coves but be aware of an incoming tide.
Great Ormes Head, Llandudno, Clwyd	Fossils	A huge Carboniferous limestone headland and great for fossil collecting; you can walk or take a car or tramway to the summit.

LIVING COAST

Location	Features	Description
Bardsey Island, Llŷn Peninsula	Bird watching, grey seals and cetaceans	Boat trips can be taken out to the island to see a diverse seabird colony, plus a chance to see grey seals and dolphins.
Porthdinllaen, Enlli and the Coastal Path around Porth Meudwy, Llŷn	Seals, grey	View from headlands and bays in calm seas.
Ty Cross, Rhosneigr and Porth Trecastell, Anglesey	Cetaceans	The coastline around the west coast of Anglesey has nutrient-rich water which brings in the fish – then the bigger marine life is sure to follow.
South Stack, Anglesey	Bird watching	Sea cliffs in the RSPB reserve provide nesting sites for as many as 9,000 seabirds, including puffins, guillemots, razorbills, kittiwake and fulmars.
Bull Bay, Anglesey	Cetaceans	Risso's dolphins, harbour porpoises and occasionally minke whales can be seen from headlands.
Dee Estuary, Welsh shoreline	Bird watching	Mudflats, wetlands, saltwater lagoons, coastal sand dunes, shingle and sea cliffs offer an amazingly diverse habitat for dozens of bird species.

HUMAN COAST

Location	Features	Description
Machynlleth to Pwllheli, Cardigan Bay	Railway	The Cambrian line runs around Cardigan Bay from Aberystwyth to Pwllheli and crosses the Barmouth trestle bridge; allow 2hrs 30mins.
Barmouth Bridge, Gwynedd	Bridge, rail	The wooden trestle bridge opened in 1867 and carries the Cambrian Coast railway across the wide, shallow Mawddach estuary; now listed Grade II*.
Criccieth, Llŷn Peninsula	Historic, Megalithic tomb and Norman castle	This area has been settled since the Bronze Age and a megalithic burial chamber lies east of the town; the castle was built around 1230.
Caernarfon Castle, Gwynedd	Historic, defence, Medieval	One of the great Medieval castles, re-built by the English King Edward I in the late thirteenth century during a bitter war with the Welsh princes.
Britannia Bridge, Menai Strait, Gwynedd	Bridges, road and rail	The bridge across the Menai Strait was built by Robert Stephenson in 1849 and provides a rail and road link from the mainland to Anglesey.
Caer Idris, Anglesey	Historic, defence, pre-history	An Iron Age hill fort occupies a commanding position on the summit of a ridge, overlooking the Menai Straits.
Twyn-y-Parc, Anglesey	Historic, defence, pre-history	A late Iron Age hill fort in the south-west of the island, created by strong ramparts running across the narrow neck of a long, straggling promontory.
Holyhead, Anglesey	Port	A ferry service to Dublin has been running from Holyhead since Elizabethan times; today, there are four sailings a day.
Dinas Porth Ruffydd, Treardour, Anglesey	Historic, defence, pre-history	A late Iron Age promontory fort (cliff castle) on the west coast of Anglesey; difficult access across precipitous cliffs.
South Stack Lighthouse, Anglesey	Lighthouse	Opened in 1809 to add safe passage for ships on the treacherous Dublin-Holyhead-Liverpool sea route; the light can be seen for 44km (28 miles).
Dinas Gynfor, Anglesey	Historic, defence, pre-history	A classic Iron Age promontory fort on the most northerly tip of Anglesey, overlooking the Irish sea.
Ynys-y-Fydlyn, Anglesey	Historic, defence, pre-history	This hill fort becomes an island at high water; the inhabitants had very good visibility along this stretch of coast from The Skerries to Holy Island.
Mynydd Llwydiarth, Anglesey	Historic, defence, pre-history	A late Iron Age hill fort built on a rocky spur on the east coast of Anglesey, overlooking Traeth-coch (Red Wharf Bay).
Beaumaris Castle, Anglesey	Historic, defence, Medieval	A stunning symmetrical castle overlooking the entrance to the Menai Straits, built by Edward I as part of his campaign to conquer north Wales.
Llandudno, Conwy	Railway, funicular	This cable-hauled cliff tramway runs from Llandudno to the Great Orme headland; it is the only funicular in the country to travel on public roads.
Llandudno Pier, Conwy	Pier	The pier opened in 1877 and is the longest Wales. Partially rebuilt in 2012, the structure is now listed as Grade II*.
Rhyl to Holyhead, north Wales	Railway	View the north Wales coastline from Llandudno, through Bangor, and over the Menai Strait to the island of Anglesey; journey 2hrs.

ACTIVITIES

Location	Features	Description
Wales Coast Path	Walks and coastal paths	Launched in 2012, the Wales Coast Path closely follows the entire coastline of Wales; 1,400 km (870 miles) long, allow 6 to 8 weeks.
Bangor to Aberdaron, Gwynedd	Walks and coastal paths	The Llŷn Pilgrim's Trail is an ancient route which follows the coastline and offers views of Ireland; 116km (72 miles), allow a week.
Pwllheli, Llŷn Peninsula	Sailing, windsurfing	Home to the Welsh National Sailing Academy and a major sailing centre, with a marina, dinghy sailing and regular competitions.
Abersoch, Llŷn Peninsula	Sailing, windsurfing	A popular village seaside resort with great beaches; the sailing school offers RYA-certified courses in sail, power, and jet skis.
Hell's Mouth (Porth Neigwl), Llŷn Peninsula	Surfing	A gently sloping beach facing the south-west offers relatively large waves for surfers and kayakers; there is a surf school and shop.
Porth Neigwl, Llŷn peninsula	Surfing	A vast sand and shingle beach with plenty of beach breaks, plus a small reef at the north-west end which attracts more experienced surfers.
Porth Penrthyn to Llanfairfechan to Lavan Sands, Menai Strait	Walks and coastal paths	Part of the Wales Coast Path, which offers bird watching and a nature reserve around Lavan Sands; 13km (8 miles), allow three hours.
Aberffraw to Rhosneigr, Anglesey	Surfing	These beaches face the south-west and offer good surfing, especially with a strong onshore wind; generally better conditions in winter.

MERSEYSIDE, LANCASHIRE, CUMBRIA & ISLE OF MAN

The main coastal feature in this region is Morecambe Bay, the country's biggest expanse of inter-tidal mudflats and sand covering 310sq.km (120sq. miles) – an area twice the size of the city of Manchester. These flats are the remains of an outwash plain formed by meltwaters during the last Ice Age. Today, the mudflats, salt marshes, lowland bogs and freshwater wetlands are important for overwintering birds and support rare butterflies and plants.

Further north along the Cumbria coastline are the coastal towns of Allonby, St Bees, Seascale and Silloth – all once popular Victorian seaside resorts; they all have impressive sandy beaches. There is also plenty of coastal history too. Whitehaven was originally a small fishing village but grew into the country's second busiest port in the mid-eighteenth century, exporting coal worldwide. This lovely Georgian town has been designated a 'gem town' by the Council for British Archaeology.

PHYSICAL COAST

Location	Features	Description
Sefton Coast (Formby to Southport), Merseyside	Beach, dunes, shells, driftwod	20km (12.4 miles) of sandy beaches backed by extensive shifting dunes, dune grassland and wet dune slacks (wet depressions in dune system).
Cabin Hill, Formby, Merseyside	Coastal features, shifting dunes, dune grassland, wet dune slacks	A typical beach / dune system, where lyme grass, sea rocket and sea holly to trap wind-blown sand and begin to form embryonic dunes.
Ravenmeols Sandhills, Formby, Merseyside	Coastal features, shifting dunes, dune grassland, wet dune slacks	A local nature reserve composed mainly of high mobile dune ridges with smaller embryo dunes, alongside a large and generally undisturbed beach.
Ainsdale and Birkdale Sandhills, Southport, Merseyside	Coastal features, shifting dunes, dune grassland, wet dune slacks	A natural, wild dune area and local nature reserve, comprising high dune ridges and dune valleys containing slacks (low wind-formed hollows).
River Ribble, Lancashire	Tidal flats	The River Ribble has the third largest tidal range in the country, 8m (26ft), which results in extensive sand- and mudflats.
Lytham St Anne's, Lancashire	Beach, sand and dunes	A local nature reserve, the St Anne's sand dunes can be found between the main promenade road and the beach.
Morecambe Bay, Lancashire	Tidal flats	The huge glacial outwash plain has four rivers draining into it and is the largest estuary in United Kingdom.
The Arnside Bore, Morecambe Bay	Tidal bores	The bore passes close to the town of Arnside; best in March through to September, it typically occurs two hours before high water.
Sandscale Haws, Cumbria	Beach, sand	A diverse dune habitat here supports a wealth of wildlife, and with magnificent views across the Duddon estuary to the Lakeland Fells.
Drigg Dunes, Ravenglass, Cumbria	Beach, sand and dunes	A local nature reserve, Drigg Dunes lies between one of the most attractive beaches on the Cumbria Coast, and the River Irt estuary.
St Bees Head, Cumbria	Cliffs	This red sandstone headland is one the most dramatic natural features along this coastline, with 6.5km (4 miles) of towering precipitous cliffs.
Whitehaven, Cumbria	Fossils	Whitehaven is one of the few places where you can collect Carboniferous fossil plants; well-preserved specimens can be found on the foreshore.
Grune Point, Cumbria	Coastal features, spit, shingle, salt marsh and mudflats	This 1.5km (1 mile) shingle spit in the Solway Coast Area of Natural Beauty has been created by longshore drift, and partially blocks the River Waver.

LIVING COAST

Location	Features	Description
Sefton Coast (Formby to Southport), Merseyside	Dune habitat, wildflowers, insects, reptiles	This SSSI reserve supports many dune-specialised species and recognised as one of the most important dune habitats in north-west Europe.
River Ribble, Lancashire	Salmon run	The River Ribble is a key breeding ground for the endangered Atlantic salmon.
Leighton Moss, Lancashire	Birdwatching	The largest reed bed in north-western England; amongst many others, it is home to egrets, marsh harriers, and otters.
Morcambe Bay, Lancashire	Birdwatching	Mudflats.
Walney Island, Cumbria	Seals, grey	Grey seals can be seen around the nature reserve, although there is no access into the reserve.
Isle of Man	Cetaceans	On an important migration route for basking sharks, the Isle of Man offers some of the best sightings anywhere around the British Isles.
Isle of Man: east coast from Ramsey to Port Soderick	Cetaceans	Bottlenose dolphins can be seen offshore; view from coast or boat.
Calf of Man (offshore island), Isle of Man	Seals, grey	Boat trips can be taken from the quay at Port St Mary running all year, weather permitting.

HUMAN COAST

Location	Features	Description
Liverpool, Merseyside	Harbour, port	At the beginning of the nineteenth century, 40 percent of world trade passed through Liverpool; the harbour still handles cargo and cruise ships.
Southport, Lancashire	Pier	The oldest iron pier in the country, and the second longest in England; it was significantly restored between 2000 and 2002 and is Grade II listed.
Blackpool Central Pier	Pier	One of three piers in the resort, the Central Pier was built in 1867 and is home to the funfair and the 'Big Wheel'; it suffered a major fire in 2020.
Fleetwood, Lancashire	Harbour	The town blossomed from the 1830s with a new harbour and the railway, but like many fishing ports, Fleetwood has suffered since the 1960s.
Cockersand Abbey, Thurnham, Lancashire	Religious site	Founded in 1184 as a hospital and elevated to an abbey in 1192; founded on marshland, dissolved in 1539, now listed as a Grade I building.
Barrow-in-Furness to Carlisle, Cumbria	Railway	The track follows the sea for most of the journey, and you can stop off at the seaside towns of Ravenglass, Silecroft and St Bees; allow 3hs.
Douglas, Isle of Man	Harbour	The island's only harbour handles passengers, roll-on roll-off vehicle ferries, oil and gas tankers, general cargo vessels, and fishing boats.

ACTIVITIES

Location	Features	Description
West Kirby, Wirral	Sailing / windsurfing	The Wirral Sailing Centre on West Kirby Marine Lake offers a full range of watersport activities, with courses available for all ages and levels.
Sefton Coastal Path	Walks and coastal paths	Crosby in the south to the Ribble estuary, flanked by salt marshes.
Southport, Lancashire	Sailing / windsurfing	The West Lancashire Yacht Club is based on the Marine Lake in Southport and offers safe and family-orientated sailing for sailors of all abilities.
Freckleton to Silverdale, Lancashire	Walks and coastal paths	This mostly follows the coastline between Merseyside and Cumbria with the odd diversion inland where necessary; 106km (66 miles).
Morecambe Bay, Lancashire and Cumbria	Walks and coastal paths	The area is treacherous without expert knowledge; Michael Wilson is the Queen's Guide to the Sands and will take you across safely.
Cartmel Circular, Cumbria	Walks and coastal paths	This circular walk uses part of the Cumbria Coastal Way and Cistercian Way to take you on a tour of the area surrounding this lovely coastal area.
Ulverston, Bardsea and Baycliff loop, Cumbria	Walks and coastal paths	The trail is mainly used for walking and bird watching; rated as moderate, 20km (12.5 miles).
Port St Mary, Isle of Man (south coast)	Surfing	The winter months between September and March have the best swells; the beach offers both reef and point breaks.

Niarbyl is a rocky headland on the south-west coast of the Isle of Man between Port Erin and Peel. It is a good position from which to watch for common dolphins, minke whales and basking sharks.

NORTHERN IRELAND

Much of the bedrock here is old, hard, volcanic rock, and about 70 percent of the coastline has been designated an Area of Outstanding Natural Beauty – with good reason. Not only can you visit the Giant's Causeway, but the easy Causeway Coastal Way includes Dunluce Castle and Carrick-a-Rede rope bridge. The walk will take you 2-3 days – possibly longer if you detour to the Bushmills distillery.

The coastline is very diverse, with mudflats, sandy beaches, sea cliffs, and craggy coastal outcrops which provide ideal nesting sites for gulls and other seabirds. In the inter-tidal zone you will find seagrass, crabs and small fish in rock pools, and starfish and jellyfish on the beaches. There are seals inshore, and bottlenose dolphins, porpoises and basking sharks offshore. Good viewing places include Donaghadee, Portstewart and Main Head (where you might even spot orcas). Boat trips are available from Portstewart, Portrush and Ballycastle.

PHYSICAL COAST

Location	Features	Description
Magilligan Dunes, County Derry-Londonderry	Dunes	One of the largest sand dune systems in the British Isles, this coastline is constantly changing shape; now declared a nature reserve.
Magilligan Beach, County Derry-Londonderry	Coastal feature, cuspate spit	Found at the eastern mouth of Lough Foyle, this is the best example of a beach-ridge foreland in Northern Ireland, formed after the last Ice Age.
Benone Beach, County Derry-Londonderry	Beach, sand	This Blue Flag beach offers 11km (7 miles) of sandy shoreline from Downhill to Magilligan Point and backed by sand dunes.
Downhill Beach, County Derry-Londonderry	Beach, sand	An area of outstanding natural beauty and set against a backdrop of cascading waterfalls and towering sand dunes.
Castlerock Beach, County Derry-Londonderry	Beach, sand	A Blue Flag beach, 1km (0.7 mile) long and running from the sea cliffs of Downhill to the lower River Bann estuary.
Portstewart Strand, County Derry-Londonderry	Dunes	The large dunes here extend for almost 3km (1.9 miles) from Ballyaghran Point in the west to Strandhead in the east; up to 30m (98ft) high.
Antrim coast, County Antrim	Fossils	Along the coast, sandstones and mudstones are exposed, offering fossils from the lower Jurassic period (about 200 to 180 million years ago).
West Strand, Portrush, County Antrim	Beach, sand	If you need some exercise after lying on the beach, there is also a pedestrian and cycle promenade here which follows the coastline.
Giant's Causeway, County Antrim	Rock formations	About 40,000 interlocking basalt columns formed during a volcanic intrusion 60 to 50 million years ago; a UNESCO World Heritage Site.
Whitepark Bay, Country Antrim	Beach, 'singing sand', fossils	Various fossils, including ammonites and belemnites, plus late Iron Age flint axes and arrowheads; the sands make a humming sound when you walk across.
Ballintoy Harbour Beach, County Antrim	Beach, sand	The beach offers secluded bays, rock pools and sandy coves; the harbour also featured in *Game of Thrones*, as the Pyke and Iron Islands.
Ballycastle, Portrush, County Antrim	Beach, sand	This really is a Blue Flag beach for everyone – sandcastle builders, swimmers, surfers, dog walkers, sunbathers and sand-walkers.
Fairhead Cliffs, Ballycastle, County Antrim	Cliffs	These dramatic cliffs are composed of vertical strands of dolerite, with steeply cracked walls columns reminiscent of organ pipes.
Rathlin Island, County Antrim	Beach, sand	This is the most northerly inhabited island off the Irish coast, with breathtaking views of the Causeway Coast.
Murlough Bay, County Down	Dunes	The shingle beach dunes are post glacial features, shaped by the wind and waves over the last 11,000 years; please stay on the dune boardwalks.
Strangford Lough, County Down	Fjords / loughs	The lough has at least seventy islands and many more islets (pladdies); designated Northern Ireland's first Marine Conservation Zone in 2013.
Cranfield West and Tyrella, County Down	Beach, sand	You will find a gently sloping, mainly sandy beach here at the entrance to Carlingford Lough, with the majestic Mourne Mountains as a backdrop.
Carlingford Lough, County Down	Fjords / loughs	A glacial sea fjord bordering the Republic of Ireland; the name comes from the Old Norse Kerlingfjrðr, meaning 'narrow sea inlet of the hag'.

LIVING COAST

Location	Features	Description
Portstewart, County Derry-Londonderry	Cetaceans	There are frequent sightings of dolphins off Portstewart Strand.
Portrush to Whitehead; County Antrim	Cetaceans	Bottle-nosed dolphins can be seen form the headlands.
Rathlin Island, County Antrim	Bird watching	The most northerly inhabited island off the Irish coast, and a Special Area of Conservation due to its large bird colony.
Murlough Bay and Dundrum Bay, County Down; Strangford Lough	Bird watching	A National Nature Reserve with a fragile 6,000-year-old sand dune system; good for walking and birdwatching.
Strangford Lough, Ards Peninsula, Rathlin Island, Copeland Islands	Bird watching, grey seals	Grey seals can be seen offshore, occasional shark sightings, and a wide and a diverse avian population for bird watchers.

HUMAN COAST

Location	Features	Description
Foyle Port, County Derry-Londonderry	Port	This is the United Kingdom's most westerly port, and an important northerly harbour for Northern Ireland handling bulk cargo.
Dunluce Castle, Portrush, County Antrim	Historic, castle, early modern era	One of the most picturesque and romantic Irish Castles; the ruins date mainly from the sixteenth and seventeenth centuries.
Derry / Londonderry to Portrush, County Antrim	Railway	The line mainly runs around the south-eastern shoreline of Lough Foyle; great views, allow 1hr 30mins.
Giants Causeway and Bushmills Railway, County Antrim	Railway, tourist	A 20-minute journey from Bushmills village to the Giants Causeway; runs four times a day between 11am and 2.30pm on Friday to Sunday.
Waterfoot, County Antrim	Harbour	This is one of the most sheltered harbours on the Antrim Coast and was probably used as early as the eighteenth century by the Vikings.
Larne, County Antrim	Port	Now a major ferry port, with regular crossing to Cairnyan in Scotland.
Larne, County Antrim	Harbour, pre-history	This was the site of a Bronze Age promontory fort and flints date from 6,000 BC; the harbour is now a major ferry port to Scotland.
Carrickfergus Castle, County Antrim	Historic, castle, Medieval	A fine Norman castle built in 1177 on a rocky promontory, originally almost completely surrounded by sea, overlooking Belfast Lough.
Donaghadee Harbour, County Antrim	Harbour, natural	Between 1760 and 1825, a daily ferry took couples to Scotland, as the law there allowed 'irregular marriages' by almost anybody in authority.
Donaghadee Lighthouse, County Antrim	Lighthouse	The impressive lighthouse at the entrance to Donaghadee harbour was the first Irish lighthouse to be converted to electric operation, in 1934.
Ardglass, Country Down	Harbour, natural	This natural inlet has been a fishing harbour for over 2,000 years and gives access at all states of the tide; still quite important commercially.
Kilclief Castle, County Down	Historic, castle, Medieval	This tower-house castle is on four levels and overlooks Strangford Lough; it was built between 1412 and 1441.

ACTIVITIES

Location	Features	Description
Coastal Causeway Route, County Antrim	Walks and coastal paths	From Portstewart to Ballycastle passing through the Causeway Coast Area of Outstanding Natural Beauty; 53km (33 miles), allow 2-3 days.
Causeway Coast, County Antrim	Surfing	There are no less than nine surfing beaches along the Causeway Coast, with Portrush being the centre of this surfing area.
Portrush, County Antrim	Surfing	Both West Strand and East Strand face north, which means the prevailing winds are offshore, creating good 'clean' waves.
Islandmagee, County Antrim	Walks and coastal paths	A hidden gem giving spectacular views to Muck Island and Scotland; choose from two routes – over the short clifftop, or the shoreside trail.
Belfast Lough, County Antrim	Sailing / windsurfing	A very popular sailing area, offering spectacular scenery and wildlife in a deep, east-facing, protected bay.
Strangford Lough, County Down	Sailing / windsurfing	One of Northern Ireland's most important sailing locations, offering fabulous sailing in protected water.
Dundrum Bay to Newcastle, County Down	Walks and coastal paths	A moderate walk with excellent bird watching along the edge of Dundrum Bay and a view of the Mourne Mountains; 16km (10 miles).

Dunluce Castle was inhabited at different times by the feuding McQuillan and MacDonnell clans. Built on the edge of a basalt outcrop, it is only accessible across the Carrick-a-Rede rope bridge.

WESTERN SCOTLAND & THE WESTERN ISLES

It is difficult to know where to begin when describing this area. This is a very exposed coastline, and a combination of ancient rocks, glacial erosion, powerful Atlantic waves, and literally thousands of islands make the west coast of Scotland simply stunning. As a sailing area it is second to none, although the unyielding seafloor and powerful tides makes careful navigation essential. There are also boat trips out of most tourist harbours to see a huge variety of birdlife and marine mammals.

There is also a network of ferries to get you from the mainland to a dozen or so offshore islands, each of them with their own distinctive character and history. If you want to visit the Outer Hebrides and you are short of time, there are one-hour flights from the mainland to Stornaway, Benbecula and Barra. The flights are expensive, but you get spectacular views from the small aircraft; if you fly to Barra, you will land on the only beach runway in the world.

PHYSICAL COAST

Location	Features	Description
Caerlaverock Wetland Centre, Solway Firth, Dumfries & Galloway	Salt marshes and wetlands	570 hectares (1,400 acres) of wild reserve on the north Solway coast, with barnacle geese and whooper swans in winter.
Ardwell Bay, south of Girvan, Ayrshire	Fossils	The Ordovician sandstones and mudstones here are black from weathering, so they need to be split; look for trilobites and brachiopods.
Rhinns of Galloway, Dumfries and Galloway	Coastal features, peninsula, natural harbour	This hammer-shaped peninsula is 40km (25 miles) at its widest; Loch Ryan offers a protected port for the Cairnryan Port to Belfast ferry.
Ardrossan, Ayrshire	Beach,	Two good sandy beaches near the town, and with an easy rail connection; the ferries to Arran take about 55 minutes.
Isle of Arran, Ayrshire	Beach, raised and relic cliffs	Since the Ice Age, the land has been rising due to the weight of ice being removed; this has resulted in beaches and cliffs left well above sea level.
King's Cave, Isle of Arran	Cliffs, sandstone	The largest of several shoreline caves north of Blackwaterfoot on the Isle of Arran; it is said this was where Robert Bruce was inspired by a spider.
Ailsa Craig, Firth of Clyde, Dunbartonshire	Rock formation, volcanic plug rocks	The granite on the island was traditionally mined to make curling stones; no longer inhabited, but you can get boat trips from Girvan.
Gulf of Corryvreckan, between Jura and Scarba, Argyll & Bute	Tidal feature, whirlpool / maelstrom	One of the best examples of a whirlpool anywhere in the world; a visit by boat during peak flow is not for the faint hearted.
Staffa, Isle of Mull	Coastal features, headland erosion, basalt columns and sea cave	Fingal's Cave is formed from hexagonally jointed basalt columns, similar to the Giant's Causeway in Northern Ireland and those on Ulva.
Staffa, Isle of Mull	Cliffs, intruded basalt	Hexagonal volcanic columns; the Vikings named the island stave or 'pillar island', as the cliffs reminded them of their houses built from vertical logs.
Ulva	Rock formation, hexagonal columns	Basalt columns dating from around 60 million years ago, formed at the same time as the rock formations on Staffa and the Giant's Causeway.
Isle of Tiree	Beach, sand	Stunning beaches and turquoise water; Balephuil, Balevullin, Caolas, Crossapol, Gott Bay, Sandaig, Scarinish & Vaul.
Isle of Eigg	Beach, 'singing sands'	The stunning white beach here is made from eroded quartz, and as you walk over it, you compress the grains, and they will 'whistle' or 'sing'.
Cleadale, Isle of Eigg	Cliffs and 'singing sands'	Dramatic basalt cliffs overlook the bay; the sands at Cleasdale 'sing' when you walk across the beach.
Elgol Beach, Isle of Skye	Coastal features, headland erosion, wave-cut notch	The clay-rich sandstone cliffs here have a distinctive wave-cut notch at sea level and have eroded in a distinctive honeycomb pattern.
Talisker Bay, Isle of Skye	Coastal features, headland erosion, sea stack	On the south-west of Skye, facing the Atlantic; coastal erosion of dark basalt has left a large sea stack off the headland to the south of the bay.
Claigan Coral Beach, Isle of Skye	Beach, sand	A near white beach is not actually made of coral, but of fossilised and sun-bleached algae washed in from the ocean.
Outer Hebrides	Beach, sand, dunes and machair (low grassy plain inland)	You are spoilt for choice on Lewis, Harris, North Uist, South Uist, Benbecula, Barra, Vatersay, Eriskay, Scalpay, Berneray and Grimsay.
Traighr Mhor, Isle of Barra, Outer Hebrides	Beach, sand and shells	A stunning white shell sand beach surrounded by machair (grassy plain), covered in wild primroses in spring; also the site of Barra's airport.
Luskentyre Sands, Isle of Harris, Outer Hebrides	Beach, sand	Often considered to be the UK's most beautiful beach; a photo of Luskentyre was once mistakenly used in a Thai tourist brochure.
Luskentyre Sands, Isle of Harris, Outer Hebrides	Beach, quicksand	There have been several incidents of people getting caught in quicksand despite no obvious signs, so take care when walking on the beach.
Port Ness, Isle of Lewis, Outer Hebrides	Rock formation, metamorphic	Dramatic folded metamorphic gneiss; these are the oldest rocks in Britain and at least 2.7 billion years old.
St Kilda, North Atlantic	Cliffs, igneous	The eroded remnant of a volcanic plug, once inhabited, now the UK's largest colony of puffins; 74km (46 miles) west of the Outer Hebrides.

Rockall, North Atlantic	Rock formation, volcanic plug	An uninhabitable granite islet 300km (190 miles) west of Scotland; it is important because it extends the exclusive economic zone (EEZ) of the UK.
Loch Torridon, Highland	Estuary, fjord	One of the best examples of glacial erosion in Scotland, the fjord is about 25km (15 miles) long.
Am Buachaille, Sandwood Bay, Sutherland	Coastal features, headland erosion, sea stack	Dramatic sea stacks in ancient sandstone found at the tip of the Rubh' a Bhuachaille headland, about 8km (5 miles) north of Kinlochbervie.
Sandwood Bay, Kinlochbervie, Sutherland	Coastal features, barrier beach	The subtly pink sand here comes from the erosion of local red sandstone; a very remote beach backed by dunes.
Cape Wrath, Highland	Cliffs	Sandstone and gneiss cliffs rising 281m (922ft); the highest sea cliffs on the British mainland are at Clò Mòr, 6km (4 miles) east of the headland.
Ceannabeinne Beach, Lairg, Sutherland	Beach, sand	Near white sand and clear water (albeit cold!); backed by rocky cliffs.

LIVING COAST

Location	Features	Description
Caerlaverock Wetland Centre, Solway Firth, Dumfries & Galloway	Bird watching	The Wildfowl and Wetlands Trust reserve covers 570 hectares (1,400 acres); in winter you can see whooper swans and barnacle geese.
Hebridean Whale Trail, Scotland	Cetaceans, basking sharks	The trail comprises more than 30 sites along the spectacular west coast, where you can spot dolphins, whales and sharks from the shore.
Isle of Eigg	Cetaceans, grey seals, basking sharks	The warm waters from the Gulf Stream bring lots of fish into the region, and they are followed by the predators; good for boat trips from Arisaig.
Outer Hebrides	Cetaceans and grey seals	Many species of dolphin and orcas can be seen from organised boat trips, available from Lewis, Harris, Uist, Eriskay, Berneray, Barra and Vatersay.
Loch Sgioport, South Uist, Outer Hebrides	Bird watching	The Bird of Prey Trail gives you a chance to watch white-tailed eagles and golden eagles throughout the year, especially late winter and early spring.
St Kilda, North Atlantic	Bird watching	A haven for birds, with more than 210 species; the island supports large populations of puffins, gannets, fulmars, guillemots, and many more.
Loch Gairloch, Highland	Cetaceans and birds	Whales, dolphins, porpoise, seals and many species of seabirds can be seen from boat trips, and occasionally sunfish, basking sharks and turtles.

HUMAN COAST

Location	Features	Description
Caerlaverock Castle, Dumfries & Galloway	Historic, Medieval, castle	An impressive moated triangular castle from the thirteenth century, and home to the Maxwell family; the grounds are open to public.
Stranraer, Dumfries and Galloway	Port	Originally a ferry port connecting Scotland with Belfast and Larne, but the service moved to nearby Cairnryan in 2011; now houses a new marina.
Troon, Ayrshire	Port	The ferry service to Larne no longer operates, but the harbour now services fishing boats and the export of timber.
Ardrossan, Ayrshire	Port	A commercial harbour, ferry terminal, and marina located which provides two ferry services to Arran.
Drumadoon Point, Isle of Arran	Historic, pre-history, Neolithic, Bronze Age and Iron Age	An Iron Age promontory hill fort occupied a commanding position on the headland; also standing stones from the late Neolithic and Bronze Age.
Rothesay Castle, Isle of Bute	Historic, Medieval, castle	This thirteenth century ruined castle has been described as one of the most remarkable in Scotland; open to the public all year.
Glasgow, Lanarkshire	Port	A very important shipbuilding industry grew here during the nineteenth century; there are still two yards building warships in the upper Clyde.
Oban, Argyll and Bute	Harbour, port	An important ferry port offering services to many of the outlying islands; also a busy fishing port and yacht marina.
Dunstaffnage Castle, Oban, Argyll and Bute	Historic, Medieval, castle	A thirteenth century castle guarding a strategic location; built by the MacDougall Lords of Lorn, but since held by the Campbell clan.
Skerryvore, Isle of Tiree	Lighthouse	Scotland's tallest lighthouse marks a very extensive and treacherous rocky reef offshore, 18km (11.2 miles) south-west of Tiree; it opened in 1844.
Kisimul Castle, Isle of Barra, Outer Hebrides	Historic, Medieval, castle	A sixteenth century castle built on a rocky islet in the bay, off the coast of Barra; it has its own freshwater wells and can be reached by boat.
Stornoway, Isle of Lewis, Outer Hebrides	Harbour	Offers an important sheltered refuge, and easy to enter at all state of tide; caters for ferries, fishing vessels and private boats.
Dunvegan Castle, Isle of Skye	Historic, Medieval, castle	Probably a fortified site from early times, the stone castle dates from the thirteenth century, but with later additions; seat of the MacLeod chief.
Duntulm Castle, Isle of Skye	Historic, Medieval, castle	Dates from the fourteenth century, and the seat of the chief of the MacDonald clan, who were constantly feuding with the nearby MacLeods.
Portree, Isle of Skye	Harbour	The harbour offers a deep-water anchorage with shelter from all wind directions; caters for fishing boats, private yachts and cruise ships.
Skye Bridge, Kyle of Lochalsh, Highland	Bridge, road	This privately funded bridge opened in 1995 and was said to be the most expensive in Europe; purchased by the government in 2004 and now free.

Fort William to Mallaig	Railway	The Jacobite Steam Train has been described as the greatest railway journey in the world; 135km (84 miles) of dramatic Scottish coastline.
Ardnamurchan Lighthouse, Highland	Lighthouse	The 36m (118 ft) pink granite tower was completed in 1849, but mains electricity was not installed until 1976, and automated in 1988.
Sinclair Castle, Caithness	Historic, Medieval, castle	Thought to one of the earliest seats of the Sinclair clan; there are two ruined castles here, from the fifteenth and seventeenth centuries.

ACTIVITIES

Location	Features	Description
Portpatrick to Killantringan Lighthouse, Dumfries and Galloway	Walks and coastal paths	A stunning 4.4km (2.7 mile) stroll along a rugged coastline; you can return via an inland, woodland walk.
Largs, Ayrshire	Sailing / windsurfing	Sailing and windsurfing tuition, and several yacht charter companies offering sailing on the west coast of Scotland.
Isle of Arran, Ayrshire	Walks and coastal paths	Offering a wide choice, with challenging walks across the mountainous ridges in the north, to gentler strolls in the south.
Isle of Bute, Argyll and Bute	Walks and coastal paths	This low-lying island with sandy wind-swept moorland has several walks of varying difficulty; the West Island Way is 40km (25 miles).
Westport Beach, Argyll	Surfing	Machrihanish is situated near Westport Beach is good for all levels but be aware of strong rip currents; board hire and tuition locally.
Isle of Tiree	Surfing	Anything from peeling longboard waves to barrelling beach breaks, the water is also warmed by the Atlantic Gulf Stream; local surf school.
Coffin Road on Isle of Harris, Outer Hebrides	Walks and coastal paths	You can follow the route taken by pallbearers carrying the dead for burial on the machair; 14.5km (9 miles), allow 3-4 hours, or longer with a coffin.
Isle of Lewis, Outer Hebrides	Surfing	Its exposed position to the Atlantic means the island has some of the most consistent waves in northern Europe; Gulf Stream warms the water too.
Thurso East, Highland	Surfing	Arguably the best right-hander in Scotland, but with the wind from the Arctic, you need a wetsuit with gloves, boots and a hood.

Neist Point lighthouse on the Isle of Skye in Scotland became operational in 1909. An aerial cableway is used to take supplies to the lighthouse and cottages.

EASTERN SCOTLAND & THE NORTHERN ISLES

The coastline of eastern Scotland is more protected than its western neighbour, and it is therefore less rugged – but no less spectacular. Most rocks in the region are either igneous, or deposits of Old Red Sandstone, modified by recent Ice Ages. The coastline here is packed with history; for example, the Picts living here in the ninth century were regularly attacked by the Vikings, but they successfully repelled the assaults from their formidable promontory forts, such as at Dunnottar. Many more clifftop castles were built in later centuries, including Tantallon, Keiss, and Findlater Castle at Cullen.

North of mainland Scotland are the archipelagos of Orkney and Shetland. Apart from stunning coastal scenery and a wide variety of wildlife, the islands have a history going back 5,000 years to the Neolithic Stone Age. Both Orkney and Shetland also have abundant evidence of Viking settlements dating from the ninth century.

PHYSICAL COAST

Location	Features	Description
Maiden Stack, Papa Stour, Shetland	Coastal features, cliff erosion, sea stack	So called because a tiny house is said to have been built in the fourteenth century by a local lord to preserve his daughter from the attention of men.
Holl O' Boardie Cave, Papa Stour, Shetland	Coastal features, cliff erosion, sea cave	Definitely some of the finest sea caves in the British Isles; the Holl O'Boardie is 330m (1,082ft) long and wide enough to take a small boat.
Esha Ness, mainland Shetland	Cliffs	The cliffs at Esha Ness and the surrounding area are the remnants of a violent and explosive volcano, active around 395 million years ago.
Grind of the Navir, mainland Shetland	Coastal features, cliff erosion	An unusual feature where a flat amphitheatre-like area has been eroded by waves that enter through a gate-like breach in the low cliffs.
Noss Island & Bressay Island, Shetland	Cliffs	The islands are Old Red Sandstone with some basalt intrusions; rock was quarried here for building and used all over Shetland.
Villains of Hamnavoe, mainland Shetland	Cliffs	Resistant Devonian volcanic lavas, coarse ash deposits, and the relentless pounding of waves create a dramatic headland on this unforgiving coast.
St Ninian's Isle, Shetland	Tombolo	One of the best examples of a sand tombolo in Europe, which allows you to walk to the small island from the mainland.
Fair Isle, Shetland	Cliffs	The most remote inhabited island in the UK, it lies nearly halfway between Shetland and Orkney; dramatic Old Red Sandstone cliffs on the west coast.
Brough of Birsay, mainland Orkney	Tidal feature, island	The uninhabited island is accessible on foot at low tide along a largely natural causeway, which is flooded at high tide.
St John's Head, Island of Hoy, Orkney	Cliffs	This Old Red Sandstone cliff is 335m (1,128ft) high, making it the highest vertical sea cliff in the British Isles.
Old Man of Hoy, Island of Hoy, Orkney	Coastal features, cliff erosion, sea stack	A 137m (449ft) high sea stack created by hydraulic action erosion since 1750; the stack is expected soon to collapse into the sea, leaving a stump.
Rackwick Bay, Island of Hoy, Orkney	Beach, boulders	Nestled among the Hoy hills and considered one of the most beautiful places on Orkney; the sand a boulder-strewn beach is impressive.
Dingieshowe Bay (Deerness island), mainland Orkney	Tombolo	The sand bar connects the former island of Deerness to the mainland, and you can now drive across on the A960.
Duncansby Head, John O'Groats, near Wick, Highland	Coastal features, cliff erosion, sea stacks	Three dramatic, pointed sea stacks south of Duncansby Head, about 2km (3 miles) east of John O'Groats; the cliffs are Old Red Sandstone.
Nairn Beach, Moray	Coastal features, barrier beach	An enormous sandy beach extending east towards Moray, which stretches 0.8km (0.5 miles) at low tide; dolphin sightings are fairly common.
Spey Bay, Moray	Beach, gravel	The largest shingle beach in Scotland constantly changes shape creating a variety of natural habitats, from shingle to grassland and salt marsh.
Bullers of Buchan, Aberdeenshire	Coastal features, cliff erosion, sea cave, sea arch	A collapsed sea cave forms an almost circular chasm, locally called 'the pot' some 30m (98 ft) deep, open to the sea through a natural archway.
Lunan Bay, Angus	Beach, sand and pebble	This east-facing beach is backed by sand dunes and low cliffs; traditional fishing is practised here with nets strung out to trap fish on a falling tide.
Seaton Cliffs, Arbroath, Angus	Cliffs	Spectacular Old Red Sandstone cliffs, with many sea caves, arches, stacks and blowholes; a wildlife reserve.
St Andrews and other Fife beaches, Fife	Beach, sand	St Andrews is famous for its stunning white sand beaches both north and south of the town; West Sands features in the 'Chariots of Fire' race.
Craiglaw Point, Gosford Bay, East Lothian	Fossils	Easy fossil collecting for young children; rich in Carboniferous marine fossils, including corals, bryozoans, crinoids and brachiopods.
Barns Ness, East Lothian	Fossils	There are plenty of foreshore rocks here from the Carboniferous period, 322 to 331 million years old; look out for fossilised corals and shells.
St Abb's Head, Berwickshire	Cliffs	Softer sedimentary rocks have eroded over time, leaving a high headland of harder 400-million-year-old volcanic lava.

LIVING COAST

Location	Features	Description
Shetland Islands	Cetaceans, dolphins, orcas, minke whales	A total of 23 species of whales and dolphins have been recorded in the islands, including the bluefin, sei and humpback whales.
Hermaness, Island of Unst, Shetland	Bird watching	Britain's most northerly point and National Nature Reserve (NNR); a cliff-top haven for fulmars, gulls, shags, gannets, puffins and kittiwakes.
Hoy, Orkney	Bird watching	A RSPB Nature Reserve, expect to see fulmars, puffins, red-throated divers, great skuas, hen harriers, stonechats and many more.
Hill of White Hamars, South Walls, Orkney	Seals, grey and birdwatching	This has a range of coastal features, including cliffs, caves, arches and stacks; watch seals swimming offshore and cliff-nesting seabirds.
Pentland Firth and Orkney	Cetaceans, orcas and minke whales	Migrating cetaceans pass through the straits between the mainland and Orkney Islands and can be seen from headlands and bays in calm seas.
Caithness headlands	Cetaceans, dolphins, orcas, minke whales	There are various headlands on the north-east coast which give excellent vantage points for watching a wide variety of cetaceans.
Chanonry Point, near Fortrose, Moray Firth	Cetaceans, bottlenose dolphins	With views across the Moray Firth, this narrow peninsula is an ideal location for sightings of dolphins; the best time is 1 hour after low water.
Bullers of Buchan, Aberdeenshire	Bird watching	The high cliffs offer a nesting site in spring for kittiwake, puffin, fulmar, shag, razorbill, guillemot and several species of gull.
Bass Rock, Firth of Forth, East Lothian	Bird watching and grey seals	Home to over 150,000 gannets, and many other seabird species; take a boat trip to the island and get close up to basking seals as well.
St Abb's Head, Berwickshire	Seals, grey	There was no seal colony here in 2007, but since then there has been an explosion in numbers, with 1,806 grey seal pups recorded in 2020.
St Abb's Head, Berwickshire	Bird watching	The cliff top National Nature Reserve is home to fulmar, shag, herring gull, kittiwake, puffin, guillemot and razorbill.

HUMAN COAST

Location	Features	Description
Muckle Flugga, Isle of Unst, Shetland	Lighthouse	Britain's most northerly lighthouse, built on a series of sharp rocks rising up out of the sea; built for the Royal Navy during the Crimean War.
Sullom Voe, mainland Shetland	Harbour, natural	A petroleum terminal on the north mainland is one of the biggest facilities of its kind in Europe, handling over 25 percent of the UK's oil production.
Lerwick, mainland Shetland	Harbour, natural	Norse raiders named this landfall Leir Vik, or 'Muddy Bay'; now used by fishing boats, cargo vessels, mail boats, ferries, cruise ships and yachts.
Fitful Head, Sumburgh, Shetland	Historic, modern era, defence, radar	The radar dome on here houses a secondary radar that is part of the navigational system used by trans-Atlantic flights.
Jarlshof, near Sumburgh, Shetland	Historic, pre-history, Viking, Medieval	Neolithic people first settled here around 2700 BC; discoveries include Bronze Age houses, Norse long houses and a Medieval farmstead.
Brough of Birsay, mainland Orkney	Historic, Picts, Viking, late Medieval	Rings and brooches found on the island suggest it was a Pictish power centre; later evidence of Viking houses and even a sauna.
Skara Brae, Sandwick, Orkney	Historic, pre-history	A stone Stone Age settlement on the Bay of Skaill, on the west coast of the mainland, with houses which include stone hearths, cupboards and beds.
Maeshowe, mainland Orkney	Historic, pre-history	This Stone Age chambered cairn and passage grave probably dates from around 2,800BC, and was built from huge sandstone blocks.
Kirkwall, Orkney	Harbour, natural	Said to be Orkney's best-preserved Viking town, Kirkwall was established in a natural harbour; it now serves fishing, yachts and cruise ships.
Scapa Flow, Orkney	Harbour, natural	This natural deep-water harbour was an anchorage for Viking longboats; it was a major naval base during the First and Second World Wars.
Hackness Martello Tower & Battery, South Walls, Orkney	Historic, modern era	These well-preserved military defences were built during the Napoleonic Wars, but never saw action; open to public.
Inverness to Thurso	Railway	See a wild and remote part of Scotland with vast peat bogs, salmon rivers, and whisky distilleries thrown in for good measure; allow 3hr 50mins.
Inverness to Kyle of Lochalsh	Railway	The Kyle Line runs coast to coast along Loch Ness and through the highlands; allow 2hrs 40mins.
Inverness	Harbour, natural	This deep natural harbour trades with the EU and elsewhere, and handles fuel, wind turbines, timber, animal feed, frozen fish and other goods.
Tioram Castle, Inverness, Highland	Historic, Medieval, castle	A ruin dating from the twelfth century; a principal stronghold of Clann Ruaidhrí, a powerful Medieval clan.
Fort George, Nairnshire	Historic, early modern era	After the Jacobite unrest in the mid-eighteenth century, George II built the biggest artillery fortification in Britain, if not in Europe.
Peterhead, Aberdeenshire	Port	A shipbuilding and whaling port in the eighteenth and nineteenth centuries; now a major port servicing the oil and gas industries.
Aberdeen, Aberdeenshire	Port	The traditional industries of fishing, paper making, shipbuilding and textiles have gradually been taken over by the gas and oil industry.

Location	Features	Description
Kinnaird Head Castle and Lighthouse, Aberdeenshire	Historic, Medieval and early modern era	Originally built in the 1500s, Kinnaird Head Castle was altered in 1787 to house Scotland's first lighthouse; open to public.
Dunnottar Castle, Aberdeenshire	Historic, Medieval, castle	One of the most romantic sites for a castle; the garrison held out against Cromwell's army and saved the Scottish Crown Jewels from destruction.
Dundee, Angus	Port	Thought to be established in the eleventh century, Dundee became the world centre for jute production; the harbour is now redeveloped.
Bell Rock (Inchcape) Lighthouse, Dundee, Angus	Lighthouse	Operational since 1810, this is the world's oldest surviving sea-washed lighthouse; the masonry has stood for 200 years without replacement.
St Andrews Castle, Fife	Historic, Medieval, castle	There has been a castle on this site since the twelfth century; built on a rocky promontory overlooking a small beach called Castle Sand.
Rosyth, Fife	Port	Formerly the Royal Naval Dockyard Rosyth, the yard made the final assembly of the Queen Elizabeth-class aircraft carriers for the Royal Navy.
Fife Ports, Fife	Port	The ports of Burntisland, Kirkcaldy and Methil are satellites of Forth Ports; they support the oil industry, and handle agricultural, paper and timber.
Forth Bridge, Fife & West Lothian	Bridge, railway	This cantilever railway bridge was opened in 1890 and voted Scotland's greatest man-made wonder; it is now a UNESCO World Heritage Site.
Queensferry Bridge, West Lothian	Bridge, road	Built alongside the existing Forth Road Bridge, this crossing was opened in 2017 and takes the M90 motorway across the Firth of Forth.
Blackness Castle, Linlithgow, West Lothian	Historic, Medieval, castle	With breathtaking views over the Firth of Forth, this restored castle is often referred to as 'the ship that never sailed' due to its nautical shape.
Grangemouth, Falkirk	Port	Scotland largest port handles nine million tonnes of cargo each year; the site has container, liquid and general cargo terminals.
Forth Ports, Edinburgh, West Lothian	Port	A limited company based in Edinburgh, but operating ports around the UK, including Grangemouth, Rosyth, Leith, Dundee and Tilbury.
Edinburgh to Dundee, Fife	Railway	You can cross the Forth estuary on the iconic bridge, and the track follows the coast for part of the way; allow 1hr 30mins.
Leith (Edinburgh), Midlothian	Port	A large deep-water harbour in the First of Forth, serving cruise ships, the oil industry, and general cargo.
Isle of May Priory, Firth of Forth, East Lothian	Religious site	A Benedictine community established in 1153 which relocated to the mainland in 1318; the ruins are still standing.
Tantallon Castle, East Lothian	Historic, Medieval, castle	A ruined fourteenth century castle overlooking the Firth of Forth; it was featured in the 2013 thriller, *Under the Skin*, starring Scarlett Johansson.

ACTIVITIES

Location	Features	Description
Dunnet Bay, Caithness	Surfing	Long sandy beach that picks up swells.
Moray Firth, Highland	Walks and coastal paths	The south coast and settlements here are linked by a 80km (50 miles) coastal trail from Findhorn to Cullen; generally level, allow 3 days.
Cruden Bay to Bullers of Buchan, near Peterhead, Aberdeenshire	Walks and coastal paths	This walk takes you through the Longhaven Nature Reserve; challenging in places, 13km (8 miles), allow 3 to 4 hours.
Firth of Tay, Dundee, Angus	Sailing / windsurfing	Sailing out of the marina at Dundee but be aware of shifting sandbanks in the estuary.
Fife Coastal Path, Fife	Walks and coastal paths	The path runs for 188km (117 miles) from the Firth of Tay in the north to the Firth of Forth in the south; challenging, allow a week to 10 days.
Firth of Forth, East Lothian	Sailing / windsurfing	The east-facing estuary offers protected sailing and an interesting coastline; skippered charters on luxury yachts available out of Edinburgh.
Belhaven Bay, near Dunbar, East Lothian	Surfing	One of the best surfing places in Scotland; especially good for beginners, and the shallow beach has no large rocks; local surf school.
Pease Bay, Berwickshire	Surfing	Sand and rock beach; both right- and left-hand breaks, with autumn water temperatures a surprisingly pleasant 14 degrees C.

The second Tay bridge. The first one opened in 1878 but suddenly collapsed in strong winds in 1887.

NORTHUMBERLAND, TYNE & WEAR & DURHAM

The north-east of England is an oft forgotten gem. Unlike many other parts of Britain, the coastline here has no big, glitzy seaside resorts, yet the extensive sandy beaches are some of the very best in the country. Admittedly the weather this far north can be a bit on the chilly side, but this is more than made up by the charm of traditional working harbours, and several historic castles.

Bamburgh lies within the Northumberland Coast Area of Outstanding Natural Beauty, and its beach is one of the most picturesque in the country. Close by is Bamburgh Castle, and the Holy Island of Lindesfarne. Further south, boats trips from the fishing village of Seahouses will take you to the Farne Islands, with puffins in residence from March to early July. Whitley Bay and Roker are delightful seaside towns, and the beach at Seaham is worth a visit if you collect sea glass. The area is spoilt for coastal walks, which take you to old colliery towns as well as pretty bays and headlands.

PHYSICAL COAST

Location	Features	Description
Northumberland coast	Beach, sandy	There are pristine sandy beaches all along the coast, including Alnmouth, Bamburgh, Blyth, Cresswell, Druridge, Embleton, Seahouses and Spittal.
Berwick on Tweed, Northumberland	Fossils	Dark grey Carboniferous limestones reveal corals, crinoids (starfish-like creatures) and occasionally plant remains.
Scremerston, Northumberland	Fossils	A good location for range of Carboniferous fossils, including crinoidal limestone and corals, as well as plant remains and trace fossils.
Holy Island, Northumberland	Tidal feature, causeway, tidal flats	The island is a designated Area of Outstanding Natural Beauty, with sand and mudflats; take care as the causeway is covered at high tide.
Beadnell, Northumberland	Fossils	Blocks of Carboniferous shale blocks on foreshore reveal brachiopods, corals, crinoids and trace fossils; safe for young children.
Howick, Northumberland	Fossils	In a relatively small area, you can find trilobites, corals, brachiopods, plant fossils, and trace fossils from the Carboniferous period.
Tyne & Wear coast	Beach, sandy	There are plenty more sandy beaches along the north-east coast, including Cullercoats, Long Sands, Marsden, Sandhaven, Whitburn and Whitely Bay.
Seaton Sluice, Whitley Bay, Tyne & Wear	Fossils	Fossils can be easily found here in shale blocks on the foreshore, which will need careful splitting; expect to find plenty of corals and bivalves.
Marsden Rock, Marsden, South Shields, Tyne & Wear	Coastal features, erosion, sea stacks	The limestone cliffs here are unstable and care should be taken; a large section collapsed in January 2021, fortunately without injury to anyone.
Tees Estuary	Tidal flats	A large estuary with extensive mudflats and sandbanks supporting diverse wildlife; sites such as Seal Sands are protected areas.
Durham coast	Beach, sandy and some pebbles	Many excellent sandy beaches all along the coast, including Crimdon, Easington, Seaham and Seaton Carew.
Hawthorn Hive, near Easington, Durham	Coastal features, erosion, wave-cut platform	Walk under the railway viaduct and to the beach; between Hawthorn Hive and Horden Debe are flat areas of bare rock, eroded by waves.
Durham coast (Seaham to Crimdon)	Coastal features, cliff erosion, caves, sea stacks	A varied coastline of headlands and bays, with the eroded magnesium limestone producing occasional caves and stacks.
Seaham, Durham	Beach, sea glass	The discarded glass from Victorian and Edwardian bottle works at Seaham and nearby Sunderland constantly wash up on the beaches.
Nose's Point, Seaham, Co Durham	Fossils	Look for Carboniferous shales and split it carefully to reveal plant remains; also pieces of fossilised tree and bivalves can be found.
Blast Beach, Nose's Point, Seaham, Durham	Coastal features, erosion, sea arch	Sea arches and caves; tonnes of coal waste a year was dumped, producing some of the worst coastal pollution in the world; now cleaned up.

LIVING COAST

Location	Features	Description
Northumberland coast	Cetaceans, harbour porpoises, bottle-nose and white-beaked dolphins	Best seen from a high vantage point in calm conditions; Emmanuel Head (Holy Island), Dunstanburgh Castle and Cullernose Point are good places.
Northumberland coast	Seals, grey	This is one of the most important areas in Europe for the Atlantic grey seal and they can be seen at any time of the year; Seahouse harbour is good too.
Farne Islands, Northumberland	Bird watching, grey seals	Famous for puffins and seals plus many others, including eider ducks, guillemots, kittiwakes, razorbills and shags; take a boat from Seahouses.
Marsden Rock, Tyneside	Bird watching	The steep, rocky cliffs provide ideal nesting sites for cormorants, fulmars, gannets, guillemots, herring gulls, kittiwakes, razorbills and shags.
Teesmouth National Nature Reserve, Durham	Bird watching, grey seals	Several reserves on the estuary give an opportunity to see a wide variety of waders and wildfowl; also grey seals can often be spotted.

HUMAN COAST

Location	Features	Description
Lindisfarne (Holy Island), Northumberland	Religious site	One of the earliest monasteries in Britain, established in AD 635; sacked by the Vikings in AD 793 – the first raid of its kind in the British Isles.
Bamburgh Castle, Northumberland	Historic, early Medieval	The Anglo-Saxon fort was destroyed by the Vikings, later rebuilt by the Normans and restored by William Armstrong, a Victorian industrialist.
Longstone Lighthouse, Farne Islands, Northumberland	Lighthouse	Completed in 1826 and still in use; best known for the 1838 wreck of the *Forfarshire* and the role of Grace Darling in rescuing survivors.
Dunstanburgh Castle, Embleton Bay, Northumberland	Historic, Medieval	Built by Edward II in the early fourteenth century, this large, magnificent castle is built on a headland overlooking the North Sea; now in ruins.
Blyth Battery, Northumberland	Historic, modern era	The World War One coastal artillery battery was upgraded and used in World War Two; now intact and accessible through the museum.
Tynemouth Priory & Castle, North Shields, Tyne & Wear	Historic, pre-history, early Medieval and modern era	Most likely an Iron Age cliff castle, a monastery was built in the eighth century and later a castle; in the twentieth century it became a gun battery.
Port of Tyne, near Newcastle, Tyne & Wear	Port	The second largest car export hub in the UK; also handles general and bulk cargo, a cruise terminal, and a ferry to Ijmuiden (for Amsterdam).
Souter Lighthouse, Marsden, Tyne & Wear	Lighthouse	Opened in 1871 to warn of the dangerous reefs at Whitburn Steel; now decommissioned, the lighthouse is open to the public.
Sunderland, Tyne & Wear	Port	At the mouth of the River Wear, the port handles general cargoes, timber, steel, aggregates, refined oil products, limestone and chemicals.
Durham to Berwick-upon-Tweed	Railway	Take the scenic East Coast Mainline along the coast; stop an Alnmouth or Berwick for the Coast and Castle Connection bus service.
Middlesbrough to Newcastle	Railway	Runs between Newcastle and Middlesbrough along the Durham coast; 64km (40 miles), allow up to 2hrs, depending on the time.
Hawthorn Hive, near Easington, Durham	Limestone kilns	Recently restored lime kilns in the beach; in the nineteenth and early twentieth century, limestone was burnt and spread on fields as fertiliser.

ACTIVITIES

Location	Features	Description
Northumberland coastal walks	Walks and coastal paths	Try Beadnell to Low Newton, 6.5km (4 miles), Cresswell to Warkworth. 21km (13 miles), Craster to Seahouses, 16km (10 miles) and many more.
Bamburgh Beach, Northumberland	Surfing	Some of the best surfing in the north-east, especially with northerly or easterly swells.
Beadnell Bay, Northumberland	Surfing	The best surf is in April and May, and September to November; the curved beach produces waves up to 4.5m (15ft).
Seaton Sluice, Whitley Bay, Tyne & Wear	Surfing	Best in winter and spring, especially around mid-tide; best when the wind is from the south.
Durham Coastal Path	Walks and coastal paths	From Seaham in the north to Crimdon in the south, the trail runs through stunning clifftop scenery; 18km (11 miles), moderate, allow 4-5 hours.

Dunstanburgh Castle was built in the fourteenth century on the remains of an Iron Age fort. Its strategic position was still useful during the Second World War as an observation post, protected by a minefield.

YORKSHIRE & LINCOLNSHIRE

The coastline of Yorkshire is unusual in that it comprises the same Jurassic sandstones and Cretaceous chalk that is found along the south coast. For this reason, places like Flamborough Head look very much like the Old Harry rocks of Dorset. Danes Dyke near Bridlington is the place to go for fossilised sea urchins, starfish and sponges. Yorkshire is famous too for some lovely seaside towns like Scarborough, Whitby, Staithes, Bridlington, Filey and Saltburn (famous for its cliff railway).

Further south into Lincolnshire, the coast flattens out and the rolling hills comprise chalk, limestone and sandstone. Skegness, Cleethorpes and Mablethorpe are traditional seaside resorts, and the coastline offers miles of unspoilt golden sands. Further south the shoreline smooths more into the Wash. This huge bay of mudflats and sandbanks is fed by the rivers Witham, Welland, Nene and Great Ouse. The Wash is now a Site of Special Scientific Interest and a National Nature Reserve.

PHYSICAL COAST

Location	Features	Description
Millclose Howles, near Redcar, North Yorkshire	Fossils	You can easily collect loose Jurassic fossils washed up on the shingle foreshore; a good, safe site for children.
Saltburn, North Yorkshire	Beach, sandy, Jurassic fossils	The beach runs 6km (3.8 miles) to the River Tees; good amenities, clean sand; the foreshore offers ammonites, belemnites and brachiopods.
Saltburn, North Yorkshire	Cliff erosion	Huntcliff is a dramatic vertical sea cliff rising 111m (365ft) from the shoreline; it lies about 1.5km (1 mile) east of the town.
Staithes to South Speeton, North Yorkshire	Cliff erosion	A spectacular stretch of coastline about 67km (42 miles) long comprising sand, limestone and mud cliffs, subject to increasing cliff erosion.
Port Mulgrave, North Yorkshire	Fossils	A wide range of ammonites, mostly found in nodules; best place is in the middle of the bay, where fossils can be seen at eye level after a storm.
Runswick Bay, near Whitby, North Yorkshire	Beach, sand, fossils	A pretty village leads down to a small sandy beach; good for rock pools, coastal walks and Jurassic fossils (including ammonites).
Whitby, North Yorkshire	Beach, sand	West Cliff is the larger beach with typical 'bucket and spade' attractions; the smaller Tate Hill beach is more sheltered and allows dogs all year.
Filey, North Yorkshire	Beach, sand	A glorious 8km (5 mile) stretch of golden sand; less commercial than many neighbouring resorts, and ideal for a quiet holiday with young children.
Flamborough Head, East Yorkshire	Cliff erosion, caves, arches, stacks plus sand / pebbly beach	Rugged chalk cliffs, with good examples of cliff erosion, including sea caves, arches and sea stacks; offers a good habitat for seabirds.
Green Stacks, Flamborough Head, East Yorkshire	Coastal erosion, arch, wave-cut platform	Classic coastal chalk erosion on the headland, with fine examples of sea arches, stacks and other erosional features.
Bridlington, East Yorkshire	Beach, sand	Good, wide, gently sloping sandy beaches run as far south as the Humber; the beach gets busy in summer, so go south to find a quieter section.
Hornsea, East Yorkshire	Beach, sand and shingle	Offers a traditional seaside resort with a promenade and landscaped gardens; dog friendly, good for walking, swimming and watersports.
Spurn Head, Holderness, East Yorkshire	Coastal feature, recurved spit	A striking 5.6km (3.5 mile) curving spit of sand and shingle banks stabilised by marram grass and sea buckthorn; caused by southerly longshore drift.
Saltfleetby Theddlethorpe Dunes, Lincolnshire	Sand dunes	An extensive and important reserve covering 556 hectares (1,374 acres); includes tidal sand- and mudflats, salt and freshwater marshes and dunes.

LIVING COAST

Location	Features	Description
Bempton Cliffs, East Yorkshire	Bird watching	Around half a million seabirds roost in the chalk cliffs between March and October; an RSPB reserve.
Flamborough Head, East Yorkshire	Bird watching	An important seabird colony with thousands of birds; expect to see fulmar, guillemots, kittiwake, puffins, razorbills, and autumn migration.
Flamborough Head, East Yorkshire	Cetaceans, dolphins, porpoises and whales	Bottlenose dolphins and harbour porpoises are frequently seen offshore, and occasionally minke and humpback whales.
Spurn Head, Holderness. East Yorkshire	Bird watching, wildflowers	The observatory here is a great place to watch migrating birds, which are funnelled down the narrow spit of land; also good for wildflowers.
Blacktoft Sands, East Yorkshire	Bird watching	Part of the upper Humber estuary; the extensive tidal flats provide an excellent habitat for avocet, bitterns, harriers and migrating birds.
River Great Ouse, East Yorkshire	Bird watching	The tidal flats and water meadows offer a great habitat for wintering birds, wildfowl and raptors.
Gibraltar Point, Lincolnshire	Bird watching	An unspoilt and dynamic coastline running from Skegness to the Wash; very diverse wildlife including terns, waders and common seals.
The Wash, Lincolnshire	Bird watching, common seals	The Wash National Nature Reserve is the biggest of its kind in England, covering over 89sq.km (34sq. miles) and an important wetland habitat.
Donna Nook National Nature Reserve, Lincolnshire	Seals, grey	See grey seals when they give birth in November and December along 10km (6.2 miles) of coast between Grainthorpe Haven and Saltfleet.

HUMAN COAST

Location	Features	Description
Saltburn, North Yorkshire	Seaside resort, pier	A delightful small Victorian town, with water-balanced cliff lifts, and the last remaining pier in Yorkshire.
Saltwick Bay, North Yorkshire	Historic, wreck	The *Admiral Von Tromp* trawler was wrecked on the beach in October 1976 with loss of life, in mysterious and unexplained circumstances.
Whitby Abbey, Whitby, North Yorkshire	Religious site	Opened in AD 657, this early Christian monastery later became a Benedictine abbey; on the East Cliff overlooking the North Sea.
Robin Hood's Bay, North Yorkshire	Seaside resort, fossils	A pretty fishing village on the Heritage Coast with a lovely family and dog-friendly beach, rock pools and Jurassic fossils (ammonites and bivalves).
Scarborough, North Yorkshire	Seaside resort	The town sits on a horseshoe-shaped bay; a lovely location and sandy beaches have made this a firm favourite for over 400 years.
Scarborough Castle, North Yorkshire	Historic, defence, Medieval	A wooden castle was built here in the 1130s, upgraded to stone from the 1150s with more additions over the centuries; a ruin since the Civil War.
Chalk Tower, Flamborough Head, East Yorkshire	Lighthouse	The current lighthouse here dates from 1806; a visitor centre tells the history of this important navigational beacon.
Hornsea, East Yorkshire	Seaside resort	A traditional resort with a sandy / shingle beach, landscaped gardens, a promenade; dog-friendly, good for walking, swimming and watersports.
Spurn Lighthouse, East Yorkshire	Lighthouse	There was a lighthouse from at least 1427; since the seventeenth century, a pair of lighthouses formed leading lights to guide vessels to the estuary.
Haile Sand Fort & Bull Sand Fort, Humber Estuary, East Yorkshire	Historic, defence, modern era	Both forts guarded the Humber estuary; built during the First World War and upgraded in time for the second; now given a Grade II listing.
Hull, East Yorkshire	Port	Marine trading started here as early as the thirteenth century; now thriving as a container, cruise and ferry terminal, and dry docks.
Humber Bridge, East Yorkshire and Lincolnshire	Bridge, road	This single-span road suspension bridge opened in June 1981; at the time, it was the longest bridge of its kind at 2.2km (1.38 miles) long.
Grimsby, Lincolnshire	Port	On the south bank of the Humber estuary, the docks date from the 1790s; once a thriving fishing harbour but has suffered from declining catches.
Ingoldmells, Lincolnshire	Seaside resort	A sandy beach, several caravan parks and a Fantasy Island theme park; good for swimming and walking; dogs allowed on part of the beach all year.
Skegness, Lincolnshire	Seaside resort	Busy in peak season, with a good, sandy beach; holiday camp, caravan parks, pleasure beach, donkey rides, crazy golf, crabbing or rock pooling.

ACTIVITIES

Location	Features	Description
Saltburn, North Yorkshire	Surfing	Offers some of the best and most challenging reef breaks to be found in England; lessons and board hire also available.
Scarborough, North Yorkshire	Surfing	The centre of surfing in Yorkshire, with three beaches suitable for all levels of surfers, with surf schools and surf shops.
Helmsley to Filey Brigg, North Yorkshire	Walks and coastal paths	The Cleveland Way, 169km (105 miles) long, combines heather moorland with dramatic coastline; rated medium, allow 9+ days.
Bridlington to Spurn Head, East Yorkshire	Walks and coastal paths	A coastal walk along mainly sand and shingle terrain, and some low cliffs; moderately easy, 66km (41 miles) long, allow 2-3 days.
Bridlington Bay, East Yorkshire	Sailing / windsurfing	The traditional home of the Yorkshire Coble, a small wooden fishing boat; these pretty, red-sailed sailing boats have seen a revival in recent years.
Mablethorpe to Skegness, Lincolnshire	Walks and coastal paths	This 27km (17 miles) section of the England Coast Path opened to the public in February 2019; mainly a beach walk, rated easy, allow a long day.
Skegness, Lincolnshire	Sailing / windsurfing	A wide variety of watersports on offer, including canoeing, kayaking, sailing, swimming and even beginner's surfing.
Wash Coastal Path, Boston in Lincolnshire to Kings Lynn, Norfolk	Walks and coastal paths	A largely level trek through some of England's most remote coastal scenery; 75km (47 miles), rated medium, allow 3 days.

NORFOLK, SUFFOLK & ESSEX

Most of East Anglia has widespread surface sediments of relatively recent origin, deposited in the past 2 to 3 million years and re-worked during the Ice Ages and interglacial periods. These sands, clays and peat deposits were laid down in rivers, swamps and marshes that bordered an ancient North Sea. This accounts for the gently rolling hills, low coastline and very high rates of coastal erosion in many areas. However, there are also lovely sandy beaches, and widespread mudflats and salt marshes further south in Suffolk and Essex.

There are very pleasant coastal villages and resorts here, such as Hunstanton, Wells, Blakeney, Cromer, Aldeburgh and Orford. East Anglia also has some of the most important bird watching sites in the country; seals are also frequently seen. There are plenty of examples of coastal defences too, from eighteenth century Martello towers to pillboxes and other Second World War structures.

PHYSICAL COAST

Location	Features	Description
Norfolk beaches	Beach, sand backed by dunes, some mud and shingle	There is almost a continuous beach running for 120km (75 miles) from Hunstanton to Great Yarmouth, supporting several seaside resorts.
Hunstanton, Norfolk	Cliffs, red and white striped	This purpose-built seaside town is famous for its striped cliffs, formed from different sandstone and pebbly sandstone deposits being exposed.
Hunstanton, Norfolk	Fossils, Cretaceous	A wide variety of fossils from 135 to 70 million years old; ammonites, belemnites, brachiopods, coral, echinoids, and maybe shark's teeth too.
Hunstanton Beach, Norfolk	Beach, sand, rock pools, occasional fossils	The only west-facing beach in East Anglia, making it relatively sheltered from sea breezes and good for watching the sunset; no dogs.
Brancaster Staithe, Norfolk	Salt marshes and wetlands	The sediment deposits from longshore drift have created a protected area of salt marsh and tidal flats behind Brancaster Beach.
Scolt Head, Burnham Norton, Norfolk	Coastal feature, barrier island, shingle, sand, dunes, salt marsh, mudflats	A barrier island between Brancaster and Wells-next-the-Sea; a ferry from Burnham Overy Staithe runs mid-July to early September.
Blakeney, Norfolk	Coastal feature, curved sandy spit	Longshore currents flowing east to west cause the River Glaven to divert westwards; a similar, smaller feature exists nearby at Brancaster Staithe.
Cley and Salthouse Marshes, Norfolk	Salt marshes and wetlands	This large area of wetlands was purchased in 1926 as a bird-breeding sanctuary and proved to be a blueprint for nature conservation.
Cromer, Norfolk	Coastal feature, glacial ridge and cliff erosion	A large ridge was laid down across north Norfolk as outwash from glacial streams after the last Ice Age; cliffs now quickly eroding around Cromer.
Cromer to Winterton-on-Sea, Norfolk	Cliff erosion, sand beaches, dunes	35km (22 miles) of coastline comprising cliffs, sand dunes and beaches; wave action and tidal currents are causing significant coastal erosion.
Pakefield, Suffolk	Fossils, Cretaceous and more recent	You can find almost anything here, including ammonites, belemnites, echinoids, reptiles (including ichthyosaurs), mammals and birds.
Amber coast, Southwold to Felixstowe, Suffolk	Fossils, amber	Amber is fossilised tree resin; it is easily missed as raw amber looks like a dull, brown stone, and only becomes a beautiful gemstone when polished.
Minsmere, Suffolk	Salt marshes and wetlands	This Special Area of Conservation, 8km (5 miles) south of Southwold, covers 23sq.km (9sq. miles) of tidal mudflats, salt marsh, and reed beds.
Orford Ness, Suffolk	Coastal feature, sand and shingle cuspate foreland shingle spit	A southerly-flowing longshore current carries sand and shingle down the coast, diverting the River Ore; the spit is about 16km (10 miles) long.
River Deben, Suffolk	Coastal feature, sand and shingle spit	A southerly-flowing longshore current blocks the entrance to the River Deben with shingle, making this a difficult entrance in a small boat.
River Deben, Suffolk	Salt marshes and wetlands	Despite a narrow entrance into the North Sea, the river opens out upstream into a large area of tidal flats; important for overwintering birds.
Walton-on-the-Naze, Essex	Cliff erosion	It is estimated that the low, sandy cliffs here are eroding at an average rate of about 1m a year, but this could increase with sea level rise.
Walton-on-the-Naze, Essex	Fossils, Cretaceous and more recent	A very unpredictable site, best after a storm; famous for fossil bird remains, large shark's teeth and plant remains.
Blackwater Estuary, Essex	Salt marshes and wetlands	A large wetland area between Maldon and West Mersea; oysters have been harvested here since Roman times.
St Osyth, Essex	Coastal feature, sand and mud spit	A classic current-formed sandy spit on the east bank of the River Colne, encloses an area of tidal creeks (Ray Creek) and mudflats.
Dengie, Bradwell-on-Sea, Essex	Coastal feature, chénier	Low shelly beach ridges separated by muddy deposits; this area is highly mobile during onshore storms.
Maplin Sands, Foulness Island, Essex	Beach, quicksand	A large area of mudflats on the north bank of the Thames estuary; in the 1970s, it was proposed as the site of a new London airport.

LIVING COAST

Location	Features	Description
Norfolk coast	Cetaceans	Harbour porpoises seen offshore and in the Blackwater estuary; very rare sightings of sperm, minke and Sowerby's beaked whales.
Titchwell Marsh, Norfolk	Bird watching	Diverse habitats including reed beds, salt marsh and freshwater lagoons, with nesting avocets and marsh harriers; an RSPB reserve.
Scolt Head Island, between Brancaster and Wells, Norfolk	Bird watching	A large breeding colony of sandwich terns, plus little terns, migratory waders and wintering wildfowl; summer ferry from Burnham Overy Staithe.
Cley Marshes, Norfolk	Bird watching	Cley and Salthouse Marshes are run by the Norfolk Wildlife Trust; six hides available with views across managed pools and scrapes.
Blakeney Point, Norfolk	Seals, grey	Home to the largest grey seal colony in England; you cannot walk out to the point, but boat trips operate from Morston and Blakeney in summer.
Minsmere, Suffolk	Bird watching	An RSPB Nature Reserve includes reed beds, lowland heath, acid grassland, wet grassland, woodland and shingle vegetation; wide variety of birdlife.

HUMAN COAST

Location	Features	Description
Norfolk coast	Seaside resorts	The visitor is spoilt for choice with resorts boasting wide, sandy beaches: Hunstanton, Sheringham, Cromer and Great Yarmouth.
Brancaster, Norfolk	Historic, Roman	Branodunum dates from the third century, one of 11 Saxon shore forts on the south and east coasts, used to protect shipping and repel raiders.
Wells Next-The-Sea, Norfolk	Harbour	There has been a harbour here for more than 600 years, and it is still used by commercial and fishing vessels, as well as private craft.
Norwich to Sheringham, Norfolk	Railway	Combine East Anglia's rural landscape with stops at two of the area's most popular holiday resorts – Cromer and Sheringham; allow 1hr.
Cromer Pier, Norfolk	Pier	There was a jetty here as long ago as 1391, but the current cast iron pier opened in 1901; it was badly damaged by a storm surge in 2013.
Happisburgh, Norfolk	Lighthouse	Built in 1790, this is the oldest working lighthouse in East Anglia, and the only independently run lighthouse in Great Britain.
Burgh Castle, Great Yarmouth, Norfolk	Historic, Roman	Burgh Castle was another Saxon shore fort, built to house troops as a defence against Saxon raids up the rivers of the east coast.
Southwold, Suffolk	Pier	Built in 1900 for steamships bringing tourists from London, Clacton and Great Yarmouth; now restored with a pub, theatre and function room.
Southwold, Suffolk	Lighthouse	An important navigational aid for vessels navigating the east coast, and guides vessels heading for Southwold harbour; open to the public.
Dunwich Village, Suffolk	Lost settlement, Medieval	The town was once the early Medieval capital of East Anglia, but now mostly lost to the sea after eight centuries of coastal erosion.
Greyfriars, Dunwich, Suffolk	Religious site	A Franciscan friary founded before 1277 was soon threatened by coastal erosion and completely moved inland in 1289; now a ruin, listed Grade II*.
Martello Tower, Aldeburgh, Suffolk	Historic, nineteenth century	Built between 1808 and 1812 to protect against a possible French attack, the tower is owned by the Landmark Trust; can be rented for vacations.
Admiralty Scaffolding, Bawdsey Beach, Suffolk	Historic, Modern era	Steel pipes driven into a sandy beach to deter invasion forces during the Second World War; once very common along the south and east coasts.
Sutton Hoo, Woodbridge, Suffolk	Historic, Medieval burial mound	One of the most important archaeological discoveries in Britain, this is the site of a grave, burial ship and artefacts of an Anglo-Saxon king.
Tide Mill, Woodbridge, Suffolk	Historic, tidal feature	The earliest record of a tide mill here goes back to 1170, and it was the last in the country to close in 1957; now restored and open to the public.
Martello Towers, Felixstowe, Suffolk	Historic, nineteenth century	Four Martello towers (designated P, Q, T and U), built to protect the entrances to the rivers Deben and Orwell, all within a 5km (3.1 mile) stroll.
Felixstowe Docks, Suffolk	Harbour	Our busiest container port, handling 48 percent of the UK's container traffic, with two terminals and a Roll-On Roll-Off facility.
Orwell Bridge, Ipswich, Suffolk	Bridge, road	This concrete box girder bridge opened in 1982 and carries the A14 to Felixstowe docks; the bridge is closed by high winds several times a year.
Cudmore Grove Battery, East Mersea, Essex	Historic, Modern era	Situated in Cudmore Grove Country Park (a nature reserve), is a Second World War 4.7-inch gun emplacement and observation post.
Manningtree to Harwich, Essex	Railway	The Mayflower Line runs along the Stour estuary and through an Area of Outstanding Natural Beauty; 18km (11 miles), allow 20 min.
Bradwell-on-Sea, Essex	Historic, Roman	Othonae was built at the entrance to the rivers Blackwater and Colne, the latter which led to the important city of Camulodunum (now Colchester).
Southend-on-Sea, Essex	Pier	Extending 2.1km (1.3 miles) out into the Thames estuary, Southend boasts the longest pleasure pier in the world.
Tilbury Docks, Thurrock, Essex	Harbour	Part of the wider Port of London, Tilbury has extensive facilities for containers, grain, bulk cargoes, vehicle imports and a cruise terminal.

ACTIVITIES

Location	Features	Description
North Norfolk Harbours	Sailing / windsurfing	Blakeney, Brancaster, Burnham Overy and Wells have sheltered harbours and launching slips, making them ideal for all kinds of watersports.
Hunstanton to Cromer, Norfolk	Walks and coastal paths	The Norfolk Coast Path passes through wide beaches, sand dunes, nature reserves and lots of birds; 72km (45 miles), easy / moderate, 3-4 days.
Lowestoft to Felixstowe, Suffolk	Walks and coastal paths	The Suffolk Coastal Path follows the coastline and passes through coastal heath, 98km (61 miles), graded easy, allow 3-4 days.
Suffolk Estuaries	Sailing / windsurfing	The rivers Alde, Deben, Orwell and Stour are all protected estuaries with excellent facilities for all types of boat, several launching slips available.
Essex Rivers	Sailing / windsurfing	The Blackwater, Colne, Crouch are the main rivers in Essex, all are muddy and shallow, with sailing clubs and marinas catering for all tastes.
Cudmore Grove, East Mersea, Essex	Walks and coastal paths	A delightful short walk, taking in marshland, oyster fisheries, old forts and 300,000-year-old mammal bones, 6km (3.7 miles), easy, allow 3 hours.
Canvey Wick to Canvey seafront, Essex	Walks and coastal paths	The Canvey Island Circular Walk is a 22km (13.7 miles) loop trail, with a mix of residential, industrial and bird watching, graded easy, allow a day.

Shoeburyness East Beach at low water. The remains of a defence boom can be seen on the horizon; it was built in 1944 to prevent enemy shipping and submarines from entering the River Thames.

ACKNOWLEDGEMENTS & CREDITS

The author and publisher would like to thank the numerous people who have contributed to this book:

- Tom Heap who kindly wrote the Foreword.
- The publisher's Oxford University contemporaries who read the sections relating to their expertise: Patrick Bird (Geology), Chris Foster (Geography) and Heather Gilmore (Biology), although we would stress that any mistakes and inaccuracies in the final book are not theirs.
- The numerous photographers who make their fantastic work available through iStock, Pixabay, Shutterstock, Wikipedia and other channels (see below).

All photographs and diagrams are © Peter Firstbrook except the photographs detailed below where the copyright holder is identified. Every effort has been made to identify and correctly attribute the copyright holders. Please advise of any errors which will be corrected in any subsequent editions of this book.

iStock

_jure (P114, 120 bottom middle, 123 bottom right)
aaprophoto (P155 bottom right)
Acceleratorhams (P40 bottom left)
AGAMI stock (P145 top left)
AlexeyMasliy (P113 bottom)
AndyRowland (P54 top)
AnjoKanFotographie (P123 top middle)
AnnaDudek (P191 bottom)
Arousa (P102 middle right, 119 bottom right, 124 top right)
aspas (P144)
birdsonline (P10 left, 155 left, 159 left)
Catuncia (P104 bottom left)
Cheezzzers (P90 bottom left)
Crazylegs14 (P84 left)
Damocean (P110 left)
Dan Olsen (P110 right)
Davemhuntphotography (P131 bottom left)
daverhead (P201 top)
Denja1 (P138 top middle)
diverroy (P212)
edb3_16 (P105)
EMFA16 (P102 middle middle)
espy3008 (P20 bottom)
Gerald Corsi (P157 top)
Hans Hillewaert (P116 top right)
Harald Schmidt (P89)
Hhelene (P120 bottom left)
Hija (P47)
JG1153 (P148 top)
Johnandersonphoto (P113 top left)
jon666 (P192 top left)
kathywj (P65)
Kazakov (P125)
KevinJCook (P43 bottom)
Lemanieh (P198 left)
LesleyJacques (P176 top)
MagicBones (P124 top left)
MajaCvetojeric (P122 top right)
Marcin_Kadziolka (P88 bottom)
Marcouliana (P127 bottom right)
Mark Stephan (P85 top)
Martin Voeller (P139 top left)
Mgodden (P70)
Michel VILARD (P140, 158 top)
Miguel Angelo Silva (P138 top left)
Mike Batson (P207 top right)
MikeLane45 (P163 bottom left)

miles_around (P58 top)
Moha El-Jaw (P91 middle right)
monitor6 (P195 bottom)
naumoid (P174)
Neil_Burton (P166)
Partha Bhowmick (P109 top)
Peter Llewellyn (P79)
PHILMACDPHOTOGRAPHY (P38-39)
Picture Partners (P122 bottom right)
Rebecca-Belleni-Photography (P135)
Richard-P-Long (P80)
Rightdx (P126 bottom)
Rob Atherton (P142 top right)
RobertMayne (P177 bottom right)
S-A-J (P128 left)
SamBarnesPhoto (P88 middle)
schrempf2 (P155 top right)
Scott O'Neill (P120 top middle)
Scubaluna (P115 bottom)
Sergio Hanquet (P149 top)
Shams (P112 bottom right)
Slowmotiongli (P120 top left, 121 left, 123 top left)
Southpict (P139 top right)
steffbennet (P159 right)
stockinasia (P187)
Sudip Biswas (P91 bottom right)
Tommy Svensson (P164)
Tsvibrav (P96-97)
tswinner (P143 bottom)
Tycson1 (P90 top middle)
urbanbuzz (P208 top)
Veliferum (P138 top right)
Vojce (P111 bottom, 115 top, 137 bottom)
Weisschr (P131 top)
wrangel (P109 bottom, 137 top, 138 bottom middle, 142 bottom)
Yerfdog (P87 bottom)
y-studio (P101 top)

Pixabay

12019 (P143 top right)
1681551 (P133 bottom, 190)
27707 (P85 bottom, 210 top)
3157171 (P9)
aitoff (P34 top left, 37)
alanhancock (P60 top)
alistairjryoung (P200 bottom right)
AndrewExtra (P195 top)
AndrewSonghurst (P63)

Arthur-ASCII (P61 bottom)
baerbel_n (P112 bottom left)
Ben_Kerckx (P62 top, 86 top)
Bergadder (P163 top)
Bluesnap (P44 left)
Cairomoon (P34 top right)
cedlambert (P177 left)
CerysGH (P102 bottom right)
Chorengel (P192 bottom)
chris_1974 (P180 right)
Christian58 (P119 top)
Collie581 (P178 left)
DenisDoukhan (P158 bottom)
DerWeg (P161 top right)
diego_perez (P69)
diego_torres (P4, 29 top, 95, 194 top)
dimitisvetsikas1969 (P30)
DorineFrequin (P161 left)
drhelenmkay (P243)
Elsemargriet (P162 left)
enriquelopezgarre (P32)
Eric Tanghe (P117)
Excellent_Tom (P192 top right)
FrankWinkler (P28 top)
Frayxx (P86 bottom)
Free-Photos (P81, 148 bottom left, 185, back cover)
Gabbe Ragio (P153)
Georg_Wietschorke (P116 bottom left, 160 bottom right)
Gillymacca (P189 left)
Giorgio_Pili_Gippi (P248)
Glucosala (P127 bottom left)
glypty1966 (P41 bottom)
Hans (P84 top right, 90 bottom right, 116 top & middle left)
HarryJBurgess (P11 top)
housedoctor3 (P31)
hunnyjar (P235)
IainPoole (P165)
ifinnsson (P157 bottom)
indianabones (P58 bottom)
InspiredImages (P34 bottom, 48-49, 49 top, 193, 207 bottom, 209)
j4sonp (P24 top)
jamaicamu (P241)
jandenouden (P152 bottom)
jLasWilson (P214)
joestrakerphotography (P238)
Jon57 (P20 top, 156 bottom, back cover)

jsbaw7160 (P121 right)
Jstolp (P152 top)
KarstenBergmann (P36 left)
Keleweasel (P51)
Ketchupbrause (P107 bottom right)
Lbradxx (P93 bottom right)
Lirinya (P60 bottom)
LoggaWiggler (P43 top left)
luxstorm (P211 bottom right)
manfredrichter (P6)
marcntomsmum0 (P151 right)
Mark_Taylor66 (P170)
Matchbox_Marketing (P222 top)
mentrea (P83)
MIH83 (P124 bottom right)
Mikes-Photography (P216-217)
MissEJB (P222 bottom)
Myagi (P200 top right)
nidan (P167)
Northern_Pinkie (P11 bottom)
Noutch (P151 left)
osvaldito66 (P73)
paulbr75 (P55, 211 top left)
Peacenik (P233)
Peggychoucair (P50)
Peter Ohlig (P227)
PeterFroehner (P160 bottom left)
platinumportfolio (P26 left)
PublicDomainPictures (P94 bottom)
Puffystriker (P210 bottom)
purple_sparrow_art (P139 bottom)
richtoze (P194 bottom)
rihaij (P163 bottom right)
Ron Porter (P189 top right)
Ronile (P148 bottom right)
sanremo-domains (P91 top right)
Sarangib (P127 top)
Shedon (P87 top)
Shutterbug75 (P28 bottom)
Siegella (P120 top right)
Stevebidmead (P176 bottom)
sujo-foto (P101 bottom)
Tama66 (P16-17, back cover)
TerriC (P112 top)
TfbWeb (P92, bottom)
TheOtherKev (P72 top)
TimHill (front cover, P13, 88 top, 168-169, 198-199)
tomas_workman0 (P29 bottom)
VANESSA_BMS (P67 top)
vinsky2002 (P161 bottom right)
Walks with angels (P18)
werdspate (P145 bottom left)
WikiImages (P128 right)
wilhei (P26 right)
Ylloh (P103)
Yorkshirepudding (P35)
zalazaksunca (P203)

Shutterstock
2Dvisualize (P124 bottom left)
Adrian Baker (P94 middle)
Billy Watkins (P184 top)
Chris Moody (P102 bottom left)
coxy58 (P119 bottom left)
Dan Bagur (P126 top)
Elena Arrigo (P44 right)

Gonzalo Buzonni (P19 right)
J Need (P123 top right)
LABETAA Andre (P129 bottom)
Oszibusz (P175 top)
Peter Turner Photography (P172)
RogerMechan (P182 top left)
Simon J Beer (P182 right)
SunFreez (P179 top left)
Wild_and_free_naturephoto (P149 bottom left)

Wikipedia
Adrian Pingstone (P84 bottom right)
Anthony Anthony (P179 bottom left)
Bart Braun (P133 top)
Bernard Finnigan Gribble (P213 bottom)
Charles de Lacy (P189 bottom right)
Charles Joseph Staniland (P213 top)
Colin Faulkingham (P104 top)
Detroit Publishing Co (P205 left)
Francis Bedford (P178 right)
Great Lakes Fishery Commission (P141 top right)
Henry Roberts (P200 top left)
Isaac Sailmaker (P200 top middle)
Islands in the Sea 2002, NOAA/OER (P108 bottom)
Jan Delsing (P120 bottom right)
John R. Dolan, NOAA (P98, 100 top right)
LPHOT Keith Morgan/MOD (P184 bottom)
Matt Wilson/Jay Clark, NOAA (P104 bottom right)
Matthias Buschmann (M.Buschmann) (P116 bottom right)
P200 bottom left
Rijksmuseum (P196)
Shane Anderson, NOAA (P107 top left, 108 top)
Umbongo91 (P77)

Creative Commons CC0 1.0 Universal Public Domain Dedication (https://creativecommons.org/publicdomain/zero/1.0/deed.en)
Bj.Schoenmakers (P141 bottom right)
British Library (P188 top)
Poliphilo (P43 top right)
W. Carter (P102 top middle & right, middle left)

Creative Commons Attribution 2.0 Generic License (https://creativecommons.org/licenses/by/2.0/deed.en)
Alek Kladnik (P107 top right)
Bruce Anderson (P100 bottom)
Charles Miller (P175 bottom)
Charlie Cars (P215)
Damien du Toit (P143 top left)
Gail Hampshire (P102 top left)
James St John (P91 middle top)
Jim Lynwood (P41 top)
Judy Gallagher (P162 right)
Phillip Capper (P40 top)
Robert Linsdell (P177 top right)
S. Rae (P122 top left, 123 bottom)
Thomas Bresson (P131 bottom right)
Tony Wood (P156 top)

Creative Commons Attribution 3.0 Unported (https://creativecommons.org/licenses/by/3.0/deed.en)
Arnstein Rønning (P130 left)
David Perez (DPC) (P141 left)
Editor5807 (P206 bottom)
Georges Jansoone (P134)
H. Crisp (P102 bottom middle, 129 top)
Hannes Grobe (P100 top left)
JoJan (P138 bottom left)
Matthieu Sontag (P131 middle right)
Stig Nygaard (P107 bottom left)
Tanya Dedyukhina (P173)
Tony in Devon (P179 bottom right)

Creative Commons Attribution 4.0 International (https://creativecommons.org/licenses/by/4.0/deed.en)
Auckland War Memorial Museum (P122 bottom left)
Bernard Picton (P111 top, 113 top right)
Chris Taklis (P142 top left)
Julien Renoult (P136)
Marie-Lan Taÿ Pamart (P160 top)

GNU Free Documentation License (https://commons.wikimedia.org/wiki/Commons:GNU_Free_Documentation_License,_version_1.2)
Matthieu Sontag (P131 middle right)

Others
Adam Li, NOAA (P146 bottom)
Alf Alderson (P211 top right)
Derek Aslett (P181)
Dr Brandon Southall, NOAA (P147, 149 bottom right)
Dr Mridula Srinivasan, NOAA (P145 top right)
Ellen O'Donnell Deerfield Community School NH, NOAA (P148 middle right)
Heather Gilmore (P86 middle)
Jeremy Atkins (P59, 61 top left, 119 middle, 132 top right)
London Array Limited (P15)
Mark Fishwick (P53 bottom)
Mr Grey (P93 left)
NASA (P54 bottom, 90 3rd right)
NOAA (P145 bottom right)
Safehaven Marine (P71)
Sealand Aerial Photography (P40 right, 52, 53 top, 56, 191 top, 225, back cover)
Simon Bassett (P211 bottom left)
The Library of Congress (P206 top)
Thomas Hawkins (P94 top)
Tony Greville (P182 bottom left)
UK Ministry of Defence (P67 bottom)
Wayne Hoggard, NOAA (P146 top)
William Daniell (P19 left)

https://creativecommons.org/licenses/by-nd/2.0/
Alan Piper (P78)

INDEX

The index covers mentions in Parts 1-3, so also look at Part 4 for specific places. If an item is in bold, there is a specific section in the book devoted to this.